Advanced
Medicinal Chemistry
A Laboratory Guide

Advanced
Medicinal Chemistry
A Laboratory Guide

Dr. M. Raghu Prasad

Professor and Head
Department of Pharmaceutical Chemistry,
Shri Vishnu College of Pharmacy, Vishnupur, Bhimavaram – 534202
West Godavari Dist., A.P., India.

Dr. A. Raghuram Rao

Professor and Head
Department of Pharmaceutical Chemistry,
University College of Pharmaceutical Sciences,
Kakatiya University Warangal – A.P., India.

PharmaMed Press
An imprint of Pharma Book Syndicate
A Unit of BSP Books Pvt. Ltd.
4-4-316, Giriraj Lane,
Sultan Bazar, Hyderabad - 500 095.

Published by

PharmaMed Press
An imprint of Pharma Book Syndicate

A Unit of BSP Books Pvt. Ltd.

4-4-316, Giriraj Lane, Sultan Bazar, Hyderabad - 500 095.

Phone: 040-23445605, 23445688; Fax: 91+40-23445611

E-mail: info@pharmamedpress.com

ISBN: 978-93-85433-01-6 (HB)

PREFACE

The man is the most affected species by diseases than any other animal species. The utility of medicine for the treatment of diseases is as old as mankind. Long before the beginning of Christian era, plants were used medicinally in India, China, Egypt and Greece. In nineteenth and early twentieth centuries the technique of extraction of active constituents from plant sources was mastered and used in the form of crude extracts. Later on, the active constituents responsible for activity were identified. Isolation of the same in the pure form and characterization was carried out. These isolated, pure, active constituents had increased the efficacy dramatically and surprisingly decreased the side effects. These observations paved a way for further continuation of research in the field and many active constituents coined as drugs were isolated and used to treat diseases.

Though these drugs were effective in curing the disorders but could not be produced in sufficient quantities to meet the market demand due to its very low yields. Thus some of the chemists involved in isolation, started synthesizing these active constituents in the laboratory utilizing the basic chemistry knowledge in an efficient and economical way. These efforts resulted in synthesizing many of the active molecules to meet the market demand.

The thirst of the man to know how the drugs act, prompted him to explore its probable mechanism of action. His work in this direction made him understand the fact that these molecules interact with the specialized structures, receptors present in the body to produce the desired physiological effect. Then he started modifying the molecules to increase the interaction with the receptor so as to get the pharmacological effect to the maximum extent possible and to have favorable pharmacokinetic profile. Thus it opened up a new era of medicinal chemistry.

The drugs synthesized were formulated into different forms like tablets, capsules, liquid orals, parentrals, ointments etc. To evaluate the amount of drug present in the given formulation many analytical procedures were developed. The stringent regulatory mechanism was framed to control the distribution of quality drugs to the customer.

In conclusion we would like to state that the nature is a rich treasure of knowledge that acts as a source for development of new drugs which are formulated into various forms and analyzed for its content.

In total medicinal chemistry includes all the aspects from isolation and characterization of active ingredients from various plant sources, synthesis of medicinal agents developed by modifying the natural products and developing analytical methods to know the amount of active constituents in the formulations.

Accordingly this book of practical medicinal chemistry has been presented in seven divisions.

Part A: Dealing with isolation of various active constituents.

Part B and C: Dealing with purification and drying of solvents and synthesis of medicinally important compounds.

Part D: Dealing with analytical methods for quantification of active ingredients in the formulation.

Part E: Dealing with analytical methods for quantification of oils and fats.

Followed by the above Part F and Part G deals with spectral work shop and bioinformatics and CADD practicals.

- Authors

ACKNOWLEDGEMENTS

Dr. M. R. P. is thankful to Sri Vishnu Raju Garu, Chairman, Sri Vishnu Educational Society, Hyderabad for his timely support for continuation of research and constant encouragement which has given the boosting for writing this book. Dr ARR thanks all his colleagues in the pharmaceutical chemistry division for necessary help. The help received from Prof. Renu Chadda and Ms. Alka Bali is gratefully acknowledged. Authors profusely thank Prof. Dr. A. V. N. Appa Rao for his valuable suggestions in Isolation of natural products of medicinal interest. The unstinted moral support received from Prof. M. C. Prabhakara is kindly acknowledged. Authors thank Ms. Sowmya Prasad, Smt. Nirmala, Mr. Devarakonda Murty, Mr. Pran Kishore, Mr. Raghav and Ms. M. Sireesha for their indispensable help during preparation of the book.

CONTENTS

ix

PART B

SOLVENT PURIFICATION TECHNIQUES

PART C

SYNTHESES OF DRUG INTERMEDIATES AND DRUG CANDIDATES (API's)

PART D

QUANTITATIVE ESTIMATION OF DRUGS / API'S IN FORMULATIONS

PART E
ANALYSIS OF OILS AND FATS

PART F
INTERPRETATION OF DIFFERENT SPECTRA

PART G
ESSENTIALS OF BIOINFORMATICS AND MOLECULAR MODELING

Part A

Isolation of Natural Products of Medicinal Interest

Isolation of Casein and Lactose from Milk

Aim: To isolate casein and lactose from milk.

Apparatus: Conical flask, funnel, stirrer, glazed tile.

Chemicals:

Chemicals and Reagents	Quantity Taken
Milk	50.0 ml
Acetic acid (10.0%)	*q.s.*
Ethanol	*q.s.*
Bromocresol green indicator	*q.s.*
Calcium carbonate	8.0 g

Principle:

Casein is a phosphoprotein present in milk and cheese. It contains at least 15 amino acids. It occurs as an amorphous white powder or granules without color or taste. It is hygroscopic in nature and present to an extent of 3% in cow`s milk.

Most proteins show minimum solubility at their isoelectric point and this property is useful to separate the proteins. In the mixture of proteins, the precipitation of specific proteins can be carried out by adjusting the pH of the solution to its isoelectric point. The isoelectric point of casein is 4.7.

Lactose monohydrate is the monohydrate of o-β-D galactopyranosyl-(1,4)-α-D glucopyranose.

3

Procedure:

Casein:

- Measure 50.0 ml of milk and transfer into a 250 ml conical flask.
- Add 10% of acetic acid drop wise with continuous stirring.
- Check the pH of the milk by using bromocresol green as an external indicator. Continue the addition of 10% acetic acid until it shows glossy green color.
- Allow the milk to stand for 10-15 minutes.
- Filter off the precipitate, wash with water several times and finally wash with ethanol.
- Dry the precipitate at room temperature.

Lactose:

- To the filtrate add 8.0 g of calcium carbonate and stir.
- Heat the mixture to boiling and filter to remove excess of calcium carbonate and proteins that coagulates after heating.
- Concentrate the filtrate to approximately 50.0 ml over a low flame.
- Add acetone or alcohol (300.0 ml) and filter.
- Cool the filtrate and collect crystals of lactose.

Chemical Tests:

- Dissolve lactose (0.25 g) in water (5.0 ml) and add ammonia (5.0 ml, 10 M solution). Heat it on a water bath for 10 minutes. A red color develops.
- Warm the mixture of hot saturated solution of lactose (5.0 ml) and dilute sodium hydroxide (5.0 ml) - Liquid becomes yellow then changes to brown. Add several drops of copper sulphate solution to the above mixture - A red precipitate of cuprous oxide is observed.

Physical Properties:

- Percentage yield: Casein: 0.5% $^W/_V$, Lactose: 0.6% $^W/_V$
- Melting point: Lactose: 203°C.

Category:

Casein:

- Nutrient.

Lactose:

- As filler and diluent in tablets and capsules.
- To prepare sugar coating solution (Sugar: Lactose :: 1:3).

References:

1. Pavia, D. L. *Introduction to Organic Lab Techniques – A Small Scale Approach*. Part –I, Ed. Brooks/Cole-Thomson Learning, 10, Davis Drive Belmont CA 94002, USA. pp 461-465.

2. Oberg, C.J. *J. Chem. Edu.* **1986**, *63*, 770.

2

Isolation of Starch from Potatoes

Aim: To isolate starch from potatoes.

Apparatus: Conical flask, funnel, stirrer, glazed tile, muslin cloth.

Chemicals:

Chemicals and Reagents	Quantity Taken
Potatoes	20.0 g
Sodium chloride solution (1.0%)	30.0 ml

Principle:

Starch is a polysaccharide obtained commercially from wheat, maize, rice and potatoes. It is present in the form of grains in different parts of plants containing amylose and amylopectin as chemical constituents.

Amylose constitutes about 20% and is made up of approximately 300 D-glucose units linked via 1-4 α glucosidic linkages. It forms a helix with about 6 glucose residues per turn. Upon action by diastase enzyme it hydrolyses completely to maltose. Acidic hydrolysis results in D-glucose. Amylopectin constitutes the major portion of starch. Diastase enzyme hydrolyses it to maltose (55%) and dextrin (45%) It has both 1-4 and 1-6 glucosidic linkages

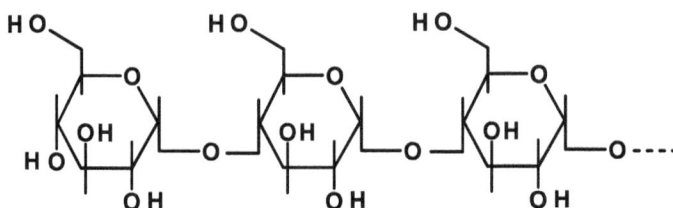

β-Amylose

6

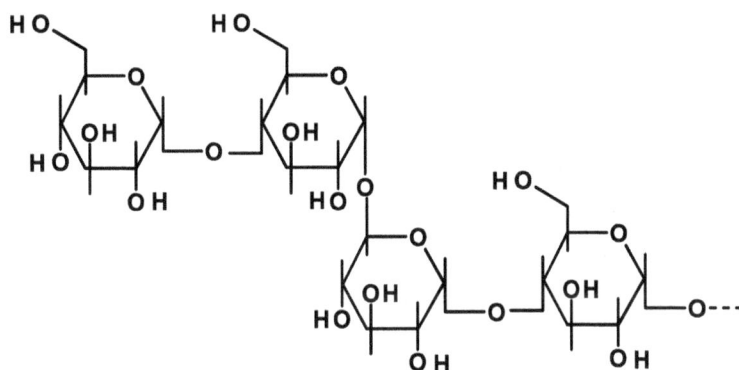

Amylopectin (α-Amylose)

Procedure:

- Wash potatoes thoroughly to remove soil and earthy matter, peel off and cut into small pieces.
- Weigh 20.0 g of potato pieces and grind with 30.0 ml of 1.0% sodium chloride (NaCl) solution in a mixer grinder and filter through a muslin cloth.
- Re-extract the residue with 10.0 ml of 1.0% NaCl solution, refilter and mix the filtrate obtained with the first filtrate.
- Allow the milky liquid to settle down and decant the supernatant liquid.
- Wash the wet starch thrice with distilled water, dry in an oven at below 60°C-65°C.

Physical Properties:

- Percentage Yield: 30% $^W/_W$

Chemical Tests:

- Take starch (1.0 g) and mix with water (15.0 ml) and cool. A translucent viscous jelly formed. Add few drops of iodine solution, the jelly turn blue which upon heating disappears and upon cooling reappears.

Category:

- Used as nutrient, demulcent, protective and absorbant.
- In preparation of dusting talcum powder.
- Antidote to iodine poisoning.
- Disintegrating agent in tablets and pills.
- Diluent in crude drugs and tablets.
- Glycerin starch is used as emolient.
- Used as base in suppositories preparation.
- Starting material for commercial production of liquid glucose, dextrose and dextran.
- Industrially used for sizing of paper and cloth.

References:

1. Raphael, I. *Natural Products – A Laboratory Guide 2ⁿᵈ Ed.* Reed Elsevier India pvt. Ltd. New Delhi, **2005**, pp 99.
2. Wallis, T. E. *Text Book of Pharmacognosy 5ᵗʰ Ed.* CBS Publishers and Distributors, New Delhi, **2002**, pp11-12.

3

Isolation of Amylose and Amylopectin from Potato Starch

Aim: To isolate amylose and amylopectin from potato starch.

Apparatus: Blender, centrifuge, stirrer.

Chemicals:

Chemicals and Reagents	Quantity Taken
Starch	10.0 g
Sodium hydroxide solution (0.16 N)	800.0 ml
Sodium chloride solution (5.0%)	230.0 ml
Sodium chloride solution (1.0%)	250.0 ml
Hydrochloric acid (1.0 N)	*q.s.*
n-Butanol	*q.s.*

Principle:

Amylose and amylopectin, the principle components of starch are separated by addition of dilute alkali followed by neutralization. Amylose remains in solution, whereas amylopectin forms a gel. The precipitation of amylose from saline solution is effected by addition of n-butanol whose removal from the complex gives pure amylose. Amylopectin can be isolated by centrifugation followed by freeze drying.

Procedure:

- Suspend freshly prepared starch without drying (10.0 g) in water (65.0 ml).
- Disperse the suspension in sodium hydroxide solution (0.16 N, 800.0 ml) with gentle stirring with the help of a glass rod at room temperature.
- Allow the alkaline solution to stand for 5 minutes and add sodium chloride solution (5.0%, 230.0 ml).

9

- Neutralize the dispersion to pH 6.5-7.5 with hydrochloric acid (1.0 N) with the aid of an external indicator like phenolphthalein.
- Allow the mixture to stand for overnight at room temperature, the amylopectin gel settles down and a clear division occurs between gel and amylose solution.
- Remove the amylose layer by suction and then filter through a filter paper.

Precipitation of amylose:

- Saturate the above filtrate by redistilled n-butanol and stir it gently on a magnetic stirrer for about 1 hour at room temperature.
- Allow the precipitated amylose-butanol complex to set for 3 hours and decant the supernatant liquid.
- Centrifuge the partially sedimented solids at 3000 rpm for 15 minutes and decant the supernatant liquid.
- Remove the butanol in the complex by passing oxygen free nitrogen through it for 15 minutes while heating in a boiling water bath.

Purification of amylopectin:

- Centrifuge the amylopectin gel by centrifuging at 8000 rpm for 20 minutes at 20°C and discard the supernatant liquid.
- Add sodium chloride solution (1.0%, 250.0 ml) to the gel with stirring and allow the mixture to stand at 18°C for 20 hours.
- Collect the purified gel by centrifuging at 8000 rpm.

Physical Properties:

- Yield: Amylose 5.0% $^W/_W$, amylopectin 14% $^W/_W$

Identification test:

- Add a few drops of iodine solution to amylose solution – Gives intense blue coloration.
- Add a few drops of iodine solution to amylopectin solution – Gives reddish coloration.

References:

1. Gilbert, L. M.; Gilbert, G. A.; Spragg, S. P. *Methods of Carbo. Chem.***1964**, *4*, 25.

4

Isolation of Calcium Citrate from Lemon Juice

Aim: To isolate calcium citrate from lemon juice.

Apparatus: Beaker, conical flask, lemon squeezer, muslin cloth.

Chemicals:

Chemicals and Reagents	Quantity Taken
Lemon juice	50.0 ml
Sodium hydroxide solution (10.0%)	*q.s.*
Calcium chloride solution (10.0%)	*q.s.*

Principle:

Citric acid is one of the commonly distributed plant acids. The lemon fruit contains about 30% of juice, which is utilized for its vitamin C content. Dried lemon peel contains about 2.5% volatile oil, hesperidin and other flavanol glycosides. Calcium chloride is added to lemon juice and citric acid is obtained in form of calcium citrate, which is carried out in basic conditions.

It is a colorless, odorless, crystalline powder, freely soluble in water and insoluble in alcohol. It looses all its water of crystallization at 120°C.

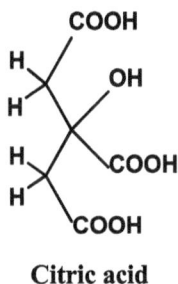

COOH

H — OH

H

H — COOH

H — COOH

Citric acid

11

Procedure:

- Take about 50.0 ml of lemon juice and add 10.0% sodium hydroxide with constant stirring until mixture is slightly alkaline as indicated by the color change from clear yellow to dark yellow.
- Strain the reaction mixture through a muslin cloth to remove the pulp.
- Again filter the filtrate through the Buchner funnel under vacuum and if the filter paper gets clogged, replace it as required for complete filtration.
- To the filtrate obtained add 5.0 ml of 10.0% calcium chloride for each of 10.0 ml of filtrate and heat the mixture to boiling.
- Filter the copious precipitate of calcium citrate while hot through a Buchner funnel.
- Wash the precipitate with small volumes of boiling water and dry.

Physical Properties:

- Percentage yield : 5.0% $^W/_V$

Category:

- It is used in production of citric acid and other citrates.
- It finds its use in improving the baking properties of flour.

References:

1. Merck Index pp 272.
2. Remingtons Pharmaceutical Sciences 19th Ed. Vol. II pp -928.
3. Kokate, C.K. *Practical Pharmacognosy* 1st Ed. pp.139.

5

Isolation of Pectin from Orange peel

Aim: To isolate pectin from orange peel.

Apparatus: Beaker, nylon cloth, heating mantle, etc.

Chemicals:

Chemicals and Reagents	Quantity Taken
Citric acid/ Tartaric acid	*q.s.*
Ammonium sulphate	*q.s.*
Ethanol /Acetone/ Propanol	600.0 ml
Orange Peels	50.0 g

Principle:

Biological source:

Orange peel is dried or fresh outer part of the pericarp of ripe or nearly ripe fruits of *Citrus aurantum* Linn. belonging to family Rutaceae.

Pectins are poly galactouronic acid esters having molecular weight 20,000 to 40,000. These polysaccharides are present in many fruits, particularly in orange and lemon rinds where they may constitute as much as 30% of the total weight.

Pectin is a complete carbohydrate compound found in nature in middle lamella of plant.

Chemically pectin is neutral methoxy ether of pectic acid. Pectins are poly uronides and consist of mixture of pectic substance like protopectin, pectin, pectic acid and calcium pectate. Pectin is a reversible colloid i.e. it may be dissolved in water, precipitated, dried and redissolved with out altering its physical properties. Pectin by addition of water forms the lumps which on heating goes into solution. Total hydrolysis of pectin yields D-galactouronic acid, methanol (CH_3OH), small amount of galactose and arabinose.

Color – cream / yellowish powder.

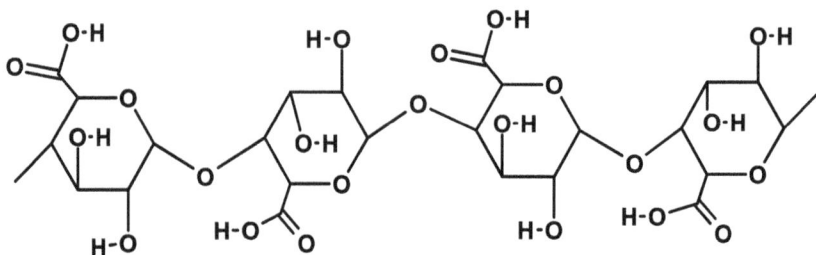

Pectin

Procedure:

- Take about 50.0 g of orange peel, cut into small pieces and wash thoroughly with water and immerse in 200.0 ml of distilled water.
- Adjust the pH to '4' with citric acid or tartaric acid or ammonium sulphate.
- Heat the mixture at 88°C-95°C for about 1 hour and filter.
- To the filtrate add 3 volumes of acidified isopropanol or ethanol or acetone and stir until pectin precipitates.
- Filter the precipitate through nylon cloth and wash for several times with 20% isopropanol or acetone to free it from acidic ions.
- Dry the product in vacuum desiccators and store in closed container.

Identification Tests:

- 10% solution when cooled form stiff gel.
- To 5.0 ml of 1.0% solution, add 1.0 ml of 2.0% solution of potassium hydroxide (KOH) and set aside at room temperature for 15 minutes. A transparent gel is formed.
- Acidify the gel with dilute hydrochloric acid (HCl) and shake well. A color less gelatinous precipitate forms which when boiled becomes white and flocculent.

Category:

- Used as an adsorbent in treating diarrhea and against constipation.
- Used as a plasma substitute.
- Used as haemostatic for internal and external hemorrhage.
- Pectin is also used in throat lozenge.
- It is used in wound healing preparations.
- Pectin is used mainly as a gelling agent, thickening agent, texturizing agent, emulsifying agent and stabilizing agent in food industry.
- Used as filler and stabilizer in confections, dairy products, fruit preparations, bakery filling, icing.
- Used as fat replacer in spreads, salad dressing, ice creams and emulsified meat products.
- It is a dietary fiber which reduces low density lipoproteins without change in high density lipoproteins and triglyceride levels.

Report:

- Yield: 60% $^w/_w$
- Melting point: 130°C.

References:

1. Miyazuki, H.; Terada, K. *Shokuhin Kogyo(In Japanese)*, **1974**, *17*, 81.
2. Takuo, S.; Minoru, O. *Appl.Environ.Microbiol.* **1980**, *39*, 908.

6

Isolation of Naringin from Grape Fruit Peel

Aim: To isolate naringin from grape fruit peel.

Apparatus: Blender, vacuum distillation assembly.

Chemicals:

Chemicals and Reagents	Quantity Taken
Grape fruit peel	50.0 g
Celite	5.0 g
Isopropanol	*q.s.*

Principle:

The albedo, the spongy inner portion of the peel, chiefly consists of cellulose, soluble carbohydrates, pectic substances, flavonoids, amino acids and vitamins. The flavonoid of the albedo is known as naringin. It was discovered by Devry in the flowers of grape fruit trees of java in 1866. It was also obtained from other sources like flowers and peel of *Citrus decumana*, grapes, Japanese bitter orange, leaves of *Pseudaegle trifoliata*.

The characteristic feature of naringin is intense bitterness. Hydrolytic enzyme, naringinase is added to grape fruit juice to convert bitter naringin into non bitter naringenin.

Naringin is extracted with hot water from the grape fruit peel along with a small quantity of pectin. Concentration of the extract to about one half with the original volume affords naringin as an octahydrate. Recrystallisation from isopropanol gives the pure dehydrate.

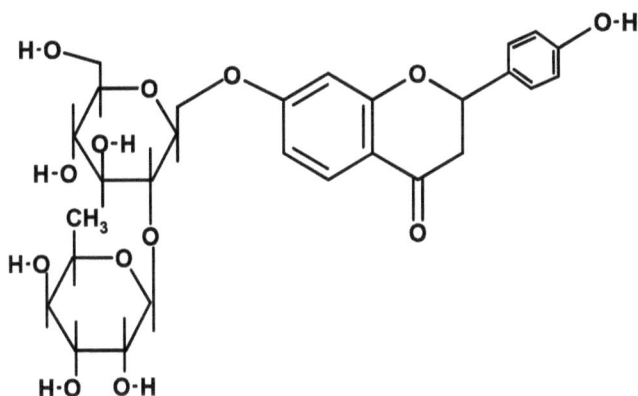

Naringin

Procedure:

- Heat 1 part chopped grape fruit peel and 4 parts water to 90°C, maintain at this temperature for 5 minutes and filter off the water extract.
- Add 2 parts of water to the solid and repeat the extraction at 80°C and filter it immediately.
- Combine the extracts and boil with 1.0% celite, filter and concentrate in vacuum to approximately one with the original volume.
- Allow the concentrated extract to crystallize in a refrigerator and then filter as an octa-hydrate of m.p. 83°C (needles).
- Dissolve naringin (8.6 g) in 100.0 ml boiling isopropanol and filter while hot.
- Heat the filtrate to boiling point to initiate crystallization then allow it to cool.
- Filter on a Buchner funnel and wash with cold isopropanol. The crude product after drying gives fine needles.
- Naringin can also be recrystallized from a small amount of hot water.

Physical Properties:

- Yield : 9.5% $^W/_W$
- Melting point: 170°C.

Category:

- Used as sweetening agent.
- Used in the synthesis of naringin dihydro chalcone.

References:

1. Poore, H. D. *Ind. Eng. Chem.* **1934**, *26*, 637.
2. Hendrickson, R.; Kesterson, J. W. *Peoc. Florida State Horticultural Soc.* **1956**, *69*, 149.

7

Isolation of Piperine from Black Pepper

Aim: To isolate piperine from black pepper.

Apparatus: Round bottom flask, reflux condenser, distillation assembly.

Chemicals:

Chemicals and Reagents	Quantity Taken
Black pepper	10.0 g
Alcoholic potassium hydroxide solution (10%)	10.0 ml
Ethanol (95%)	150.0 ml

Principle:

Biological Source:

Black pepper is dried unripe fruit of *Piper nigrum* belonging to the family Piperaceae. It also occurs in the fruit of aschanti (*Piper clusti*), long pepper (*Piper longum*) and in seeds of *Cubeba censii*. It has been isolated from *Piper famechoni* and *Piper chaba*.

The piperine content of black pepper varies from 6%-9%. In the procedure reported here, the piperine is extracted from black pepper with ethanol and converted into red trinitrobenzene complex.

Piperine is tasteless at first but it produces a burning sensation and sharp after taste. The initial tastelessness of piperine may be a consequence of its extra low solubility in water. The crude ethanol extract contains some acidic resinous material in addition to piperine and chavicine. To prevent co-precipitation of piperine and resin acids, dilute ethanolic potassium hydroxide solution is added to the concentrated extract to keep acidic material in solution or as solid gummy material that is precipitated in the vessel.

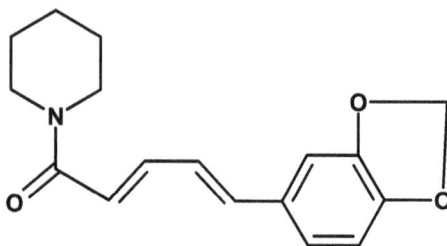

Piperine

Procedure:

- Grind 10.0 g of black pepper to a fine powder and extract with 150.0 ml of 95% ethanol in a round bottomed flask and condense for 3 hours.

- Filter the solution and concentrate in vacuum on a water bath at 60°C.

- Add alcoholic potassium hydroxide (10.0 ml &10%) to the filtrate residue and after a while decant from the insoluble residue.

- Leave the alcoholic solution over night where upon yellow needles crystallizes out.

Piperine forms a solid complex with 1, 3, 5-trinitrobenzene in a ratio1:1 in the form of red needles (melting point 130°C).

Report:

- Yield: 0.65% $^W/_W$
- Melting point: 125°C.

Category:

- Used as an anti-inflammatory agent.
- Used in the treatment of snake venom poisoning.
- Helps in absorption of Vitamin B and β-carotene.

Reference:

1. Pienemann, *Arch. Pharm.* **1896**, *243*, 204.
2. Herlant, M. *Chem. Zentr.* **1895**, *66*, 319.
3. Barille, A. *Compt. Rend.* **1902**, *134*, 1512.

8

Isolation of Caffeine from Tea Powder

(Synonym: Caffeine, thein, guarinine, methyl theobromine)

Aim: To isolate caffeine from tea powder.

Apparatus: Conical flask, round bottomed flask, funnel, beaker.

Chemicals:

Chemicals and Reagents	Quantity Taken
Tea Powder	20.0 g
Sodium carbonate	5.0 g
Celite	1.0 g
Dilute sulphuric Acid (10%)	10.0 ml
Dichloromethane	60.0 ml

Principle:

Biological Source:

Tea leaves are considered as a rich source of caffeine. Tea powder contains the prepared leaf buds of *Thea sinensis* belonging to family Theaceae.

Alkaloids have a purine nucleus. It forms a small but important group of natural products. They are often not classified as alkaloids because of their almost universal distribution in living matter and their mode of biosynthesis, which shows no relationship to amino acids from which most alkaloids arise.

Caffeine is one of the most important naturally occurring methyl derivatives of xanthine. Its concentration in variety of tea, including black and green tea, depends both on climatic and topographic conditions of

21

growth and processing methods. It is found to vary from 2.0%-4.6%. Thus, Chinese black tea contains 2.6%-3.6%, Brazilian 2.2%-2.9% and Turkish, 2.1%- 4.6% caffeine.

In addition to caffeine, other purines, theophylline and theobromine have also been reported to be present in small quantities. Michl and Haberler separated the following compounds from black tea: caffeine, 2.5% theobromine 0.17%, theophylline 0.031%, adenine, 0.014% and traces of guanine, xanthine, and hypoxanthine. Caffeine was also isolated from seeds of *Genipa americana* (2.25%) and by Stenhouse from coffee beans. *Cola nitida* is the principal source of kola nuts. It is important because of their caffeine (3%-5%) and theobromine contents.

Caffeine is used as a stimulant of the central nervous system. It has myocardial and diuretic effects and relaxes the smooth muscle of the bronchi. Caffeine is a less potent diuretic than theobromine.

The experiment involves the isolation of caffeine from tea leaves by extraction with hot sodium carbonate solution and neutralization and extraction with dichloromethane.

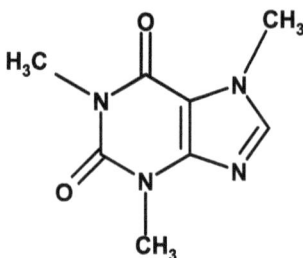

Caffeine

Procedure:

- Place the tea powder (20.0 g) in a 400 ml beaker and add sodium carbonate (5.0 g) followed by 100.0 ml of water and heat the mixture on a Bunsen flame for 20 minutes.

- Add water occasionally to keep the volume of the solution constant.

- Filter the hot solution and neutralize the filtrate cautiously with sulphuric acid (10%) while stirring.

- Then filter through a thin layer of a filter aid (celite), placed on a Buchner funnel padded with a wet filter paper and wash with dichloromethane (20.0 ml).
- Transfer the two-phase filtrate into a separating funnel and separate the organic lower layer.
- Extract the aqueous layer twice with two 40.0 ml portions of dichloromethane.
- Combine the two organic layers and distill off the solvent.
- Recrystallize the crude caffeine from a very small quantity of hot acetone or water.

Identification tests:

1. Dissolve 5.0 g of caffeine in 1.0 ml HCl and 50 mg of potassium chlorate ($KClO_3$) and evaporate on a steam bath to dryness. Invert the dish over a vessel containing ammonia.   The residue turns purple.

2. To 50.0 ml of saturated solution of caffeine add 5 drops of iodine then add 3 drops of dilute HCl. Brown colored precipitate obtained.

3. Treat few crystals of caffeine with 2-3 drops of nitric acid (HNO_3) in porcelain dish and evaporate to dryness. Add 2 drops of ammonium hydrochloride solution to residue. Purple color is seen.

Category:

- Used as stimulant of central nervous systems or psychoactive stimulant.
- It has effect on myocardium
- It has diuretic effects but less potent diuretic than theobromine.
- It relaxes the smooth muscle of bronchi.
- As analgesic in migraine.

Physical Properties:

- Yield: 1.5% $^W/_W$
- Melting point: 231°C.

References:

1. Pavia, D. L. *Introduction to Organic Lab Techniques – A Small Scale Approach.*   Part –I, Ed. Brooks/Cole-Thomson Learning, 10, Davis Drive Belmont CA 94002, USA. pp 262.
2. Michl, H.: Haberler, F. *Monatsh. Chem.* **1954**, *35*, 770.
3. Freise, F.W. *Pharm. Zentr.* **1935**, *76*, 704.
4. Stenhouse, J. *Ann.* **1854**, *89*, 244.

Isolation of Hesperidin from Orange Peel

Aim: To isolate hesperidin from orange peel.

Apparatus: Round bottomed flask, disintegrator, funnel and reflux condenser.

Chemicals:

Chemicals and Reagents	Quantity Taken
Orange peels	200.0 g
Methanol	1.0 L
Petroleum ether	1.0 L
Calcium hydroxide solution (10%)	750.0 ml
Conc. hydrochloric acid	q.s.
Dilute acetic acid	q.s.
Formamide	q.s.

Principle:

Biological Source:

Orange peel is dried or fresh outer part of the pericarp of the ripe or nearly ripe fruits of *Citrus aurantum* Linn. belonging to family Rutaceae.

Hesperidin is a non-bitter compound, predominant flavonoid in lemon and ordinary sweet oranges. Neohesperidin, (Bitter compound) an isomer of hesperidin has been found in unripe sour oranges cultivated in Europe.

It was found that the hesperidin decreases the fragility of blood capillaries.

Hesperidin can be isolated by 2 methods.

1. By extracting the dry peel successively with petroleum ether and methanol, the first solvent for removing essential oil and second the glycoside.

2. By alkaline extraction of chopped orange peel and acidification of the extract. It may be purified effectively by treatment with formamide and activated charcoal. Due to its high insolubility, crystalline nature, hesperidin is one of the easiest flavonoids to isolate.

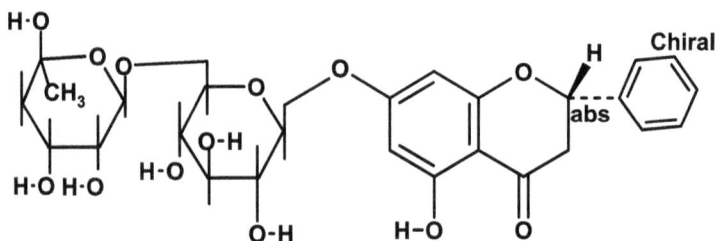

Hesperidin

Procedure:

Method A:

- Powder sun dried orange peels (200.0 g) in a disintegrator.
- Place the powder in round bottomed flask and fit it to a reflux condenser.
- Add 1.0 L of petroleum ether and heat to reflux on a water bath for 1.0 hour.
- Filter the contents of the flask while hot through a Buchner funnel and allow the powder to dry at room temperature.
- Return the dry powder into the flask, add 1.0 L of methanol and heat the contents to reflux for 3.0 hours.
- Then filter the contents while hot and wash with 200.0 ml of hot methanol.
- Concentrate the filtrate under reduced pressure which leaves a skimpy residue.
- Recrystallize from dilute acetic acid which yields white needles.

Method B:

- Take chopped orange peel (200.0g) and calcium hydroxide solution (750.0 ml, 10%) in a 2.0 L conical flask, mix thoroughly and leave it aside over night at room temperature.
- Filter the mixture through a large Buchner funnel containing a thin layer of celite on the filter paper.
- Acidify the yellow orange filtrate carefully to pH 4-5 with concentrated hydrochloric acid. Hesperidine separates as amorphous powder.
- Filter the formed precipitate with the help of a Buchner funnel, wash with water and recrystallize from aqueous formamide.
- If the precipitation of hesperidine on addition of hydrochloric acid is slow, it is advisable to concentrate the solution under reduced pressure.

Identification tests:

1. **Ferric chloride (FeCl₃) test:** Addition of $FeCl_3$ to hesperidine gives wine red color.
2. **Magnesium hydrochloride reduction test:** Drop wise addition of concentrated HCl to an ethanolic solution of hesperidine containing magnesium gives bright violet color.

Category:

- It is normally used in conjugation with ascorbic acid to minimize capillary fragility.
- It is also indicated in the preservation and management of capillary fragility or permeability is hypertension and other cardiovascular diseases.
- It is also used in habitual and threatened abortion.

Physical Properties:

- Yield: 0.05% $^W/_W$
- Melting point: 252°C.

References:

1. Leberton, P. *J. Pharm. Chim. Paris* **1828**, *14*, 377.
2. Karrer, W. *Helv. Chim. Acta.* **1949**, *32*, 714.
3. Sastry, G. P.; Row, L. R. *J. Sci. Ind. Res.* **1960**, *19B*, 500.

10

Isolation of Strychnine and Brucine from Nuxvomica Seeds

Aim: To isolate strychnine and brucine from the seeds of *Strychnos nux-vomica*.

Apparatus: Soxhlet extractor, vacuum distillation, assembly.

Chemicals:

Chemicals and Reagents	Quantity Taken
Nuxvomica Seeds	200.0 g
Calcium hydroxide suspension (10%)	200.0 ml
Chloroform	100.0 ml
Ethanol (95%)	*q.s.*
Sulphuric acid (5.0%)	*q.s.*
Sodium hydroxide (10%)	*q.s.*

Principle:

Biological source:

Nuxvomica consist of dried ripe seeds of *Strychnos nuxvomica* belonging to family loganiaceae.

Strychine and brucine are alkaloids present in *Strychnos nux-vomica*. Nux-vomica seeds contain about 3% alkaloids. Out of which a little more than half is made up of strychnine. Other alkaloids include brucine, colubrine, vomicine and pseudostrychnine. Strychnine is intensely bitter. One part of strychnine is capable of imparting bitter taste to 5,00,000 parts of water. Brucine is even bitter than strychnine. Its uses are similar to those of strychnine but its action is so much mild that it is used on a very large scale for denaturation of ethanol.

In this experiment a mixture of both is extracted with chloroform from the crushed nuts or treated overnight with calcium hydroxide. Recrystallization of crude product with 50.0% ethanol yields crystalline strychnine which is converted to its sulphate salt. Brucine is more soluble in alcohol and is recovered from mother liquor as sulphate salt.

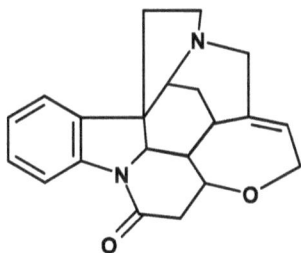   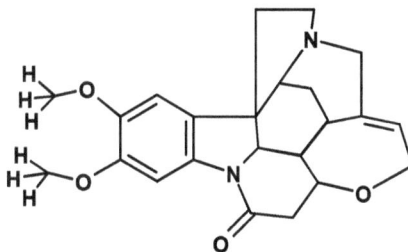

Strychnine	**Brucine**

Procedure:

- Mix thoroughly the finely grounded nuxvomica nuts (200.0 g) with suspension of 10.0% calcium hydroxide in water (200.0 ml) and leave aside for overnight at room temperature.
- After air drying, extract the slurry with chloroform (100.0 ml) in a soxholet extractor for 3.0 hours.
- Extract the chloroform layer several times with 5.0% sulphuric acid solution and subsequently basify with 10% aqueous sodium hydroxide solution. After cooling, the crystals appear.
- Filter the crystals, add 1.5 volumes of 50% ethanol and heat the mixture to reflux until most of the solid gets dissolved. Add a little amount of activated charcoal to the solution, filter while hot and leave it aside overnight. The crystals of strychnine separate out.
- Filter the crystals, wash with little amount of 50.0% ethanol. Keep the mother liquor and washings for the isolation of brucine.

Identification Tests:

- Dissolve strychnine (0.1 g) in concentrated sulphuric acid (0.5 ml) and add potassium dichromate solution. Produces deep violet color.
- Treat brucine with concentrated nitric acid. Produces yellow color.

Category:

- Due to its bitter taste nuxvomica is used as bitter stomachic and bitter tonic.
- Used as CNS stimulant.
- Strychnine increases blood pressure.
- Brucine is used as dog poison.

Physical Properties:

- Yield of strychnine and brucine: 1.45% $^W/_W$
- Melting Point of strychnine: 287°C.
- Melting Ponit of brucine: 178°C.

Reference:

1. Annet, F. A. L.; Hughes, G. K.; Ritchie, E. *Australian J. Chem.* **1953**, *6*, 58.
2. Jaminet, F. *J. Pharm. Belg.* **1953**, *8*, 339.

11

Isolation of Nicotine Picrates from Tobacco Leaves

Aim: To isolate the nicotine from tobacco powder as nicotine picrates.

Apparatus: Soxholet extractor, funnel, vacuum distillation assembly.

Chemicals:

Chemicals and Reagents	Quantity Taken
Tobacco powder	10.0 g
Alcoholic potassium hydroxide (10.0%)	QS
Diethyl ether	50.0 ml
Sulphuric acid (1.0%)	2.0 ml
Sodium hydroxide (5.0%)	*q.s.*
Picric acid	*q.s.*

Principle:

Biological source:

Tobacco consists of dried leaves of *Nicotinea tobacco* belonging to the family Solanaceae.

Nicotine is a pyridine type of alkaloid obtained from tobacco powder. Tobacco leaves contain 0.6%-0.9% of the alkaloids viz nicotine, a lesser amount of nor nicotine and aromatic principle nicotinine. In contrast to leaves, the roots of number of varieties of *Nicotinea tobacco* and other species contain at least 8 pyridine alkaloids including nicotine, nor nicotine and anabarine.

When tobacco powder is treated with potassium hydroxide, free alcoholic bases are liberated. These free bases are then extracted using solvent ether and converted to salt form using sulphuric acid and followed by basification by sodium hydroxide to liberate bases. Finally

31

the free bases are extracted with solvent ether. Nicotine being liquid alkaloid addition of picric acid, precipitates nicotine picrates from the solution. It is having yellow color.

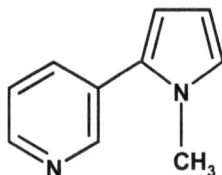

Nicotine

Procedure:

- Moisten powdered tobacco (10.0 g) with sufficient quantity of alcoholic potassium hydroxide (10.0%) to liberate alkaloid base. Dry it in an oven at below 60°C.
- Place the powder in conical flask (250 ml), add 50.0 ml of solvent ether and plug it with cotton.
- Heat on an electric water bath till it boils, shake for 2.0 min, boil again and repeat this process for 5.0 minutes. Filter and concentrate the filtrate to 20.0 ml.
- Treat the ethereal extract with 2.0 ml of 1% sulphuric acid twice.
- To the aqueous layer, add sufficient quantity of sodium hydroxide solution for basification.
- Extract the free base and concentrate. Dissolve in 3.5 ml of distilled water, filter and add saturated solution of picric acid drop wise till complete precipitate of nicotine picrate takes place.
- Keep the solution in refrigerator for half an hour, remove the supernatant liquid by decantation, dry the product and weigh.

Category:

- Nicotine exerts stimulant effects on heart and nervous system.
- It is a powerful quick acting poison.
- It is used as insecticide.

Identification tests:

- Nicotine with iodine solution in dry ether gives red crystals of iodo derivative.

Physical Properties:
- Yield: 6.0% $^W/_W$
- Melting point: 218°C.

References:
1. Jackson, K. E. *Chem. Rev.***1941**, *29*, 123.
2. Fritgsche, Reinheld. *Syracuse Chemist,* **1942**, *35,* 3.
3. Ray, M. *Rev. Intern.Tobacs.* **1948**, *23*, 132.
4. Badgett, C. O. *Ind. Eng. Chem.***1950**, *42*, 2530.
5. *Council Sci. & Ind. Res.***1953**, *45*, 666.

12

Extraction of Sennosides as Calcium Salts from Senna Leaves

Aim: To extract sennosides as calcium salts from leaves of senna by chemical method.

Apparatus: Electric shaker, vacuum unit.

Chemicals:

Chemicals and Reagents	Quantity Taken
Powdered senna leaves	50.0 g
Denatured spirit	*q.s.*
Hydrochloric acid	*q.s.*
Calcium chloride solution in ethanol (5.0%)	25.0 ml
Methanol	700.0 ml
Benzene	300.0 ml

Principle:

Biological source:

Senna powder consist of dried leaf lets of *Cassia angustifolia* belonging to family Leguminosae. It contains not less than 2.0% and the anthracene derivatives calculated as calcium sennoside-B.

Senna contains mainly 2 anthraquinone glycosides called as sennoside (A & B) which account for its purgative property. Sennosides A & B are stereoisomers of each other. They are dimeric glycosides with rhein dianthrone as aglycone. Senna leaf also contains sennosides C & D in small amounts. Besides anthraquinone glycosides senna contains 2 napthalene glycosides called tinnevilia glycoside and 6-hydroxy musizin glycoside.

34

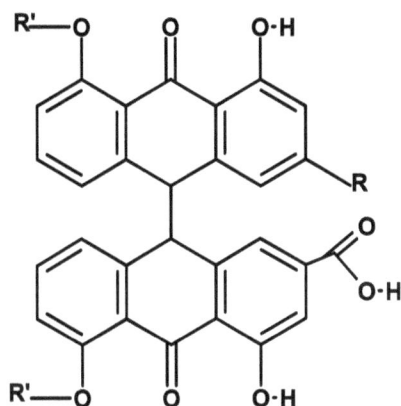

Sennoside A: R = COOH, R' = $C_6H_{11}O_5$ (Trans),

Sennoside B: R = COOH, R' = $C_6H_{11}O_5$ (Cis)

Sennoside C: R = CH_2OH, R' = $C_6H_{11}O_5$ (Trans),

Sennoside D: R = CH_2OH, R' = $C_6H_{11}O_5$ (Cis)

Procedure:

- Extract the powdered leaves of senna (50.0 g) with 300.0 ml of benzene for 2.0 hours on electric shaker.
- Then filter the solution and distill off the filtrate to separate the solvent (benzene).
- Dry the left out marc and re-extract with 70.0% methanol (300.0 ml) on shaker for 6.0 hours and filter under vacuum.
- Again reextract the marc with 400.0 ml of 70% methanol for 2.0 hours and filter.
- Combine with methanolic extracts, concentrate to $1/8^{th}$ volume and acidify to pH 3.2 by the addition of hydrochloric acid with constant stirring.
- Keep the mixture aside for 2.0 hours at 5.0°C and filter under vacuum.
- Add anhydrous calcium chloride (prepared in 25.0 ml of denatured spirit) solution (5.0%) to the filtrate with vigorous shaking, adjust the pH of the solution to 8.0 by ammonia solution and keep it aside of 2.0 hours.
- Filter the solution under vacuum and dry the precipitate over phosphorous pentoxide in a desiccator.

Identification Test:

- Borntragers test: Boil the drug with sulphuric acid and filter. To the filtrate add benzene, shake well and separate the organic layer. To the above organic layer add ammonia slowly. The ammonical layer shows pinkish red due to the presence of anthraquinone glycoside.

Category:

- It is used as a purgative by increasing the peristalsis of colon.
- It is dispensed along with carminatives to counteract the gripping effect caused by it.

Report:

Yield: 5.26% $^W/_W$

Reference:

1. Lohar, D. R.; Bhatia, R. K.; Garg, S. P.; Chawan, D. D. *Pharmacy World & Sci.***1979**, *1*, 206.

13

Isolation of Curcumin from Turmeric Powder

Aim: To isolate curcumin from turmeric powder.

Apparatus: Soxholet apparatus.

Chemicals:

Chemicals and Reagents	Quantity Taken
Turmeric powder	50.0 g
Methanol	200.0 ml
Benzene	50.0 ml
Aqueous sodium hydroxide solution (0.1%)	40.0 ml
Dil. hydrochloric acid (0.5%)	*q.s.*

Principle:

Biological source:

Turmeric consists of dried as well as fresh rhizomes of the plant *Curcuma longa* belonging to family Zingiberaceae.

Yellow coloring substances of turmeric are known as curcuminoids. The chief component of curcuminoids is known as curcumin which accounts to 50%-60%. Turmeric also contains volatile oil resin and starch.

Soxholet apparatus is used for the extraction of curcumin from turmeric powder with methanol as solvent. The extraction continued until all coloring matter is completely extracted. Methanolic extract is treated with benzene basified and acidified to precipitate extraneous matter. Then boiled to remove resinous matter and the left out solution is concentrated to get curcumin.

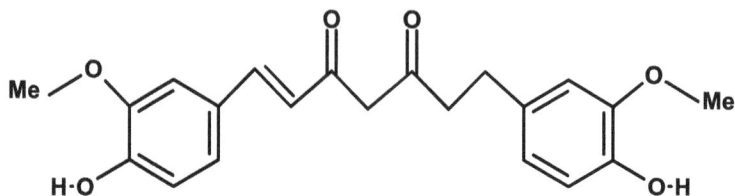

Curcumin

Procedure:

- The turmeric powder (50.0 g) was extracted with 200.0 ml of methanol for 6.0 hours in a soxholet apparatus (till all the coloring matter gets completely extracted).
- Distill off the methanolic extract to a semisolid brown colored mass.
- Dissolve the crude extract in 50.0 ml of benzene and extract twice with equal volumes of 0.1% sodium hydroxide solution.
- Combine the alkaline extracts and acidify with dilute HCl.
- Allow to settle down the yellow colored precipitate formed for about 15 minutes.
- After setting of precipitate, concentrate the extract by boiling on a water bath. During the process of boiling the resinous material agglomerates and a lumpy mass formed.
- Filter the solution in hot condition, concentrate to a very small volume and finally cool the concentrate to get the crude curcumin.

Chemical tests:

- Curcumin gives brown color with alkali (sodium hydroxide).
- Curcumin gives light yellow color with acids (HCl).

Category:

- Curcumin has been proved as anti-inflammatory drug.
- It acts as an antibacterial agent.

Physical Propertes:

Yield: 0.5% $^W/_W$

References:
1. Anderson, A. M.; Mitchell, M. S.; Mohan, R. S. *J. Chem. Edu.***2000**, *77*, 359.
2. Verghese, J. *Flavour & Fragrance J.* **1993**, *8*, 315.

14

Isolation of Trimyristin and Myristicin from Nutmeg

Aim: To isolate trimyristin and myristicin from nutmeg.

Apparatus: Reflux assembly, fractionating column.

Chemicals:

Chemicals and Reagents	Quantity Taken
Nutmeg (crushed)	30.0 g
Chloroform	200.0 ml
Ethanol (95%)	250.0 ml
Petroleum ether	250.0 ml
Activated alumina	10.0 g

Principle:

Nutmeg oil is obtained by steam distillation of the kernels of the fruit of *Myristica fragrans Houtt*, a tree 15-20 meters high, grown in Indonesia and in the West Indies. Nutmeg oil consists of approximately 90% of terpene hydrocarbons. Major components are sabinene, and α- and β-pinene. A major oxygen-containing constituent is terpinen-4-ol. A fraction consisting of phenols and aromatic ethers, such as safrole, myristicin, and elemicin, is responsible for the characteristic nutmeg odor.

Isolation of trimyristin (ester) and myristicin (phenylpropane derivative), the 2 major products of nutmeg, is accomplished by extraction with chloroform. These compounds are separated by removing the solvent and then filtering.

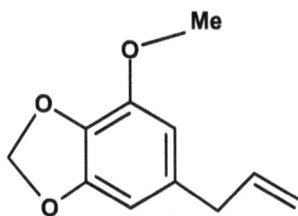

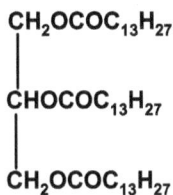

Myristicin	Trimyristin

Procedure: Isolation of trimyristin:

- Take 30.0 g of crushed nutmeg and 200.0 ml chloroform and heat it to reflux for 90 minutes on a water bath.
- Filter through a folded filter paper and dry on calcium chloride.
- Distill of the excess of chloroform under reduced pressure.
- Dissolve the left over semisolid residue in 200.0 ml ethanol (95%).
- Cool the ethanolic solution, crystalline trimyristin precipitates.
- Filter off with suction pump and wash with cold ethanol (95%). Keep the filtrate and washings for the isolation of myristicin.
- The crystals of trimyristin are colorless and odorless.

Physical Properties:

Melting Point: 54°C-55°C.

Yield: 22% $^W/_W$

Isolation of myristicin:

- Concentrate the mother liquor remaining after separation of trimyristin, a residue is obtained.
- Dissolve the residue in 20.0 ml petroleum ether and pass through a short column containing 10.0 g of activated alumina.
- Elute with 150.0 ml petroleum ether and evaporate the solvent, oil is obtained.
- Distill the oil in a fractionating coloumn.

Physical Properties:

- Boiling Point: 150°C.
- Yield: 7.0 % $^W/_W$

Category:

- Used as flavoring agent in food industry.

References:

1. Semmler, F.W. *Ber.* **1890**. *23*, 1803.
2. Power, F. B.; Salway, A. H. *J. Chem. France* **1908**, 361.
3. Verkade, P. E.; Coops, J. *Rec. Trav. Chim.* **1927**, *46*, 528.

15

Isolation of Rhein from Rhubarb Root

Aim: To isolate rhein from rhubarb root.

Apparatus: Extractor.

Chemicals:

Chemicals and Reagents	Quantity Taken
Rhubarb root	100.0 g
Water	2.25 L
Methyl isobutyl ketone	*q.s.*
Sodium bicarbonate (5.0%)	*q.s.*
Hydrochloric acid (5.0%)	*q.s.*

Principle:

Rhein was first isolated from Chinese rhubarb in 1895. It is present in various Rheum species, partly in the free and partly as a glycoside. Other natural sources of rhein are the roots of *Rumex andreaeanum* and the Brazilian species of *Cassia alata*, where it occurs mainly in a reduced glycosidic form.

   In the following experiment, rhubarb root is extracted thoroughly with water, concentrated under vacuum, and then extracted with methyl isobutyl ketone. Rhein is recovered from the latter with sodium bicarbonate followed by acidification.

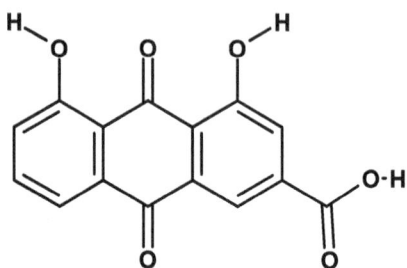

Rhein

Procedure:

- Place loosely 100.0 g of coarsely ground rhubarb root in a fluted filter paper over a wad of glass wool in an extractor.

- Extract the active material for 3 periods (1.0, 1.5 and 2.5 hours) using 750.0 ml of water for each extraction.

- Combine the extract and concentrate under reduced pressure to about 100.0 ml.

- Extract the syrupy concentrate with methyl isobutyl ketone in a continuous extractor until the extract is almost colorless.

- Take the organic solution in a separating funnel and wash with small portions (10 ml-25 ml) of 5.0% sodium hydrogen carbonate solution until the typical reddish color ceases to appear in the extracts.

- Cool the aqueous alkaline extract in ice and acidify to about pH 2 with cold dilute hydrochloric acid.

- Centrifuge the tan amorphous precipitate formed, wash with water and dry under vacuum.

- Remove the dark pigmentation by washing with acetone, followed by cold acetic acid.

- Recrystallize from acetic acid as pale yellow needles.

Physical Properties:

- Melting point: 326°C-329°C.
- Yield: 0.002% $^W/_W$

Category:

- It is bitter stomachic and intestinal astringent.
- A mild laxative.

Reference:

1. Hesse,O. *Pharm. J.* **1895**, *1*, 325.
2. Tschirch, A.; Eijken, P. A. A. F. *Chem. Zentr.* **1905**, *76*, 144.
3. Tutin. F. *J. Chem. Soc.* **1913**, *103*, 2006.

## 16

Isolation of Lycopene from Tomatoes

Aim: To isolate lycopene from tomatoes.

Apparatus: Centrifuge.

Chemicals:

Chemicals and Reagents	Quantity Taken
Tomato paste	50.0 g
Methanol	200.0 ml
Carbontetrachloride	200.0 ml
Benzene	*q.s.*

Principle:

Tomato paste is dehydrated with methanol, and lycopene is extracted from the residue with methanol-carbontetrachloride. The crude product is crystallized twice from benzene by the addition of methanol, giving lycopene of 98%-99% purity.

Further purification is achieved by a chromatographic procedure, using calcium hydroxide as the adsorbent.

Lycopene in solution undergoes isomerization even at 20°C. Crystalline lycopene is not isomerized but has a tendency to autoxidation, especially in light. It should be kept in the dark in evacuated glass tubes.

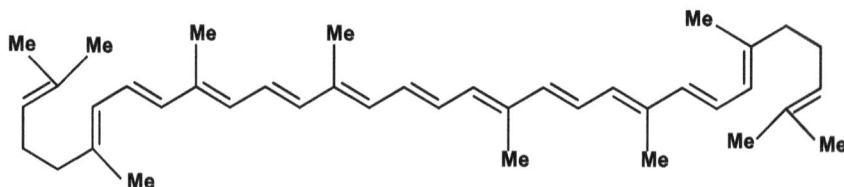

Lycopene

Procedure:

- Dehydrate the canned tomato paste (50.0 g) in a 3 L wide-mouthed bottle by adding 65.0 ml methanol. Shake the mixture vigorously and immediately to prevent the formation of hard lumps.

- Test a small sample of the suspension by hand, if it has a glutinous consistency, add more methanol to the main portion to avoid the possible clogging of filters.

- Allow the mixture to stand for 1.0-2.0 hours, shake vigorously and filter on a Buchner funnel (diameter 25-20 cm). Discard the yellow filtrate.

- Take the dark red cake into the bottle and shake with a mixture of 35.0 ml methanol and 35.0 ml carbontetrachloride. Stopper the bottle well and lift for a moment after the mixing, to release any built-up pressure.

- Repeat the brief shaking followed by opening of the bottle until no more excess pressure is noticed.

- Shake the suspension for 10-15 minutes and separate by filtration on a large Buchner funnel.

- The filtrate consists of lower very dark red carbontetrachloride phase and an orange aqueous-methanolic layer.

- Crush the slightly colored tomato residue by hand to form a nearly uniform powder and then re-extract with 35.0 ml of each solvent as described above and filter the suspension. The extraction is now almost complete.

- Combine the filtrates, transfer the methanolic layer to a 2.0 L separating funnel, and add one volume of water.

- A white emulsion appears in the upper phase. If the emulsion is reddish, stir with a glass rod until the droplets of carbontetrachloride join the lower layer.

- Separate the phases and wash the carbontetrachloride phase several times with water. Drain the carbontetrachloride solution into a 1.0 L Erlenmeyer flask and dry over anhydrous sodium sulfate.

- Pour the extract through a folded filter into a 1 L round bottomed flask equipped with a standard taper.

- Evaporate the solvent with the aid of a water pump to about 5.0 ml by heating on a water bath at 60°C.

- Transfer the solution to a similar flask of 10.0 ml capacity using a few ml of carbon tetrachloride to rinse the larger flask. Remove the solvent completely under vacuum.

- Dilute a dark oily residue left behind with a few ml of benzene and evaporate again to remove the carbontetrachloride completely.

- Transfer the partly crystalline dark residue quantitatively with 1.0 ml of benzene to a 25.0 ml Erlenmeyer flask.

- Immerse the flask in a hot water bath, a clear solution is obtained. Add boiling methanol in portions, using a dropper, to the benzene solution, with stirring after each addition, until 1.0 ml methanol has been introduced.

- Crystals of crude lycopene begin to appear immediately. Complete the crystallization by keeping the liquid first at room temperature and then in ice water.

- After standing for 1.0 to 2.0 hours, collect the crystals on a small Buchner funnel and wash with 2.0 ml of boiling methanol.

- Transfer the lycopene crystals to a tared 10.0 ml centrifuge tube and remove the last portion from the funnel with small quantities of boiling benzene.

- Add benzene to the centrifuge tube to make up the volume to 1.0 ml and dissolve the crystals by dipping the tube into hot water and stirring the contents.

- When a clear solution is obtained, add boiling methanol in small portions with a dropper, and stir the solution with a glass rod until crystals begin to appear.

- Keep the centrifuge tube at room temperature for a short time and then in an ice bath. Add more methanol in small portions, with stirring, to the cold solution such that the total volume of methanol present should not exceed 1.0 ml.

- Allow the mixture to stand for 2.0 hours in the ice bath, and separate the crystals by a brief but strong centrifuging.

- Decant the mother liquor and discard. Treat the crystals in the centrifuge tube with 1.0 ml boiling methanol.

- Stir the mixture and remove the methanol by centrifuging before it cools. Decant the methanol, and repeat the washing at least twice more.

- If the crystallization and purification is satisfactory, long, red lycopene prisms are observed under the microscope. No colorless substance should be present.
- Dry the centrifuge tube and its contents in vacuum at room temperature for a few hours and then weigh.

Physical Properties:

Yield: 0.003% $^W/_W$

Identification tests:

- Dissolve lycopene in concentrated sulphuric acid – Imparts indigo blue color.
- Dissolve small amount of lycopene in chloroform. To this add a solution of antimony trichloride in chloroform. Intense, unstable blue color produced.

Category:

- A powerful antioxidant finds it use in various cancers like prostate and mouth.
- It gaurds against ageing of skin.
- Helps in treatment of diabetis and cardiovascular diseases.
- Assists the fertility problems in men.

References:

1. Hartsen *Compt. Rend.* **1873**, *76*, 385.
2. Karrer, P.; Rubel, F.; Strong, F. M. *Helv. Chim. Acta.***1936**, *19*, 28.
3. Kuhn, R.; Bielig, H.; Dann, O. *Ber.***1940**, *73*, 1080.
4. Willstatter, R.; Esher, H. H. *Z. Physiol. Chem.***1910**, *64*, 47.

17

Isolation of Arachin and Conarachin from Groundnuts

Aim: To isolate arachin and conarachin (proteins) from groundnuts.

Apparatus: Reflux condenser, round bottomed flask, Buchner funnel.

Chemicals:

Chemicals and Reagents	Quantity Taken
Groundnuts	50.0 g
Light petroleum	200.0 ml
Sodium hydroxide Solution (0.1%)	q.s.
Sulphuric acid (5.0%)	q.s.
Sodium chloride solution (10.0%)	100.0 ml

Principle:

Groundnuts are rich in proteins. It is indicated that the total protein of groundnuts consists of globulins and a very small amount of heat-coagulable albumin. Arachin and conarachin are the 2 globulins present. Conarachin and arachin can be isolated from 2% and 10% sodium chloride respectively. Conarachin can also be isolated by complete saturation with ammonium sulphate.

Procedure:

- Take blanched ground nuts (50.0 g) in petroleum ether (200.0 ml) and heat it to reflux for 1 hour to expel the oil.
- Filter off the petroleum ether and dry the solid residue in air.
- Extract the solid residue obtained with sodium hydroxide solution (0.1%) and filter off the solution sodium proteinate.

50

- Precipitate the protein adjusting the pH to 5.0 using dilute sulphuric acid, allow it to settle. Filter and wash to free from salt and dry.
- Shake the dried protein (4.0 g) with sodium chloride solution (10.0%, 100.0 ml) to give a solution of conarachin and arachin.
- Dilute the solution with 4 times of water to precipitate the arachin.
- Filter of the precipitated arachin, wash to free from salt and dry.
- Precipitate the conarachin by heating the filtrate, filter and wash to remove the salt and dry.

Physical Properties:

- Yield: Conarachin: 10.0% $^W/_W$, Arachin: 15.0% $^W/_W$

References:

1. Jones, D.B.; Horn, M. J. *J. Agric. Res.* **1930**, *40*, 672.
2. Johns, C. O.; Jones, D. B. *J. Biol. Chem.* **1916**, *28*, 77.

## 18

Isolation of Glutamine from Beet Root

Aim: To isolate glutamine from beet root.

Apparatus: Blender, meat mincer, vacuum desiccator, vacuum distillation assembly.

Chemicals:

Chemicals and Reagents	Quantity Taken
Beet roots	1000.0 g
Ammonium sulfate solution (0.2 M)	5000.0 ml
Basic lead acetate reagent (25%)	80.0 ml
Mercuric nitrate reagent	16.0 ml
Sodium hydroxide solution (1.0 N)	*q.s.*
Sodium hydroxide solution (10.0%)	*q.s.*
Sulphuric acid (4.0 N)	0.25 ml
Ammonium hydroxide (15.0 N)	3.0 ml -5.0 ml

Principle:

Beet root is rich source of glutamine. It is also present in some other plants but these plants also contain asparagine and thus isolation is very difficult involving fractional crystallization. In beet roots the asparagines content is negligible thus this problem does not arises. It is used as an anticonvulsant agent. It is indicated in cases refractory to drug therapy.

The content of the natural glutamine in beet root is increased by treating the plant with ammonium sulphate solution. The roots are frozen and thawed to reduce the permeability of cells. The roots are grinded and extracted with cold water. The extract is treated with lead acetate solution to precipitate the interfering substances. After filtration glutamine is precipitated by adding mercuric nitrate. The precipitate formed is decomposed by passing hydrogen sulphide gas, filtered and concentrated in vacuum at low temperatures. It is recrytallized using alcohol as solvent. Mercuric nitrate is not the specific precipitant for glutamine. It also precipitates asparagines, arginine, lysine, tyrosine and cystine.

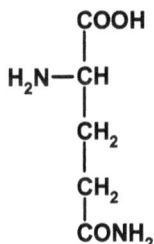

$$
\begin{array}{c}
COOH \\
| \\
H_2N-CH \\
| \\
CH_2 \\
| \\
CH_2 \\
| \\
CONH_2
\end{array}
$$

Glutamine

Procedure:

- Soak beet root (1000 g) in ammonium sulphate solution (0.2 M, 5000 ml) for 1 day, wash thoroughly and keep in cold storage.
- Cut the frozen roots and grind it in meat mincer and bring it to room temperature.
- Wrap the grinded material in a cotton cloth and squeeze, wash the material with water and squeeze it again until the volume of 1500 ml is obtained.
- Treat the effluent with basic lead acetate (25%, 80.0 ml), filter it off with the help of a buchner funnel, wash with water (600 to 800 ml) thoroughly.
- Treat the clear yellow filtrate with mercuric nitrate reagent (22.0 g red mercuric oxide + 16.0 ml concentrated nitric acid) and add water (16.0 ml).
- Heat the mixture to reflux for 4 hours (until oxide completely dissolves), cool the solution to room temperature and treat with sodium hydroxide (1.0 N) until the appearance of faint opalescence.
- Neutralize the suspension to pH 6.0 with sodium hydroxide solution (10%) and allow the precipitate to settle overnight.
- Decant the supernatant liquid, filter off the precipitate through a buchner funnel, wash thrice with water (500.0 ml).
- Suspend the precipitate in minimum amount of water, break the lumps and add sulphuric acid (4.0 N, 0.25 ml).

- Place the flask on the shaking machine and pass hydrogen sulphide gas for about 2.0 hours. Then pass the air for a short period of time to remove excess of hydrogen sulphide gas. Filter the precipitate, wash with water (200.0 ml).

- Transfer the clear, pale yellow filtrate into a 1 L round bottomed flask and concentrate in vacuum for about 20-30 minutes at 60°C bath temperature to remove last traces of hydrogen sulphide.

- Neutralize the solution to pH 6.0 by careful addition of ammonium hydroxide (15.0 N, 3-5 ml).

- Concentrate the resulting solution under vacuum at a bath temperature of 60°C-65°C to a total volume of 100.0 ml. Crystals starts appearing.

- Warm to 60°C-70°C, crystals dissolves. Add two volumes of warm alcohol and chill the flask overnight.

- Filter off the crystals, wash with 50% alcohol followed by 95% alcohol and ether, dry it in a vacuum desiccator over sulphuric acid

Physical properties:

Yield: 0.3% $^W/_W$

Category:

Anticonvulsant.

References:

1. Schulze, E.; Bosshard, E. *Ber.***1883**, *16*, 312.
2. Vickrey, H. B.; Pucher, G. W. *Biochem. Prepns.***1949**, *1*, 44.

19

Isolation of Capsanthin from Paprika

Aim: To isolate capsanthin from paprika.

Apparatus: Buchner funnel, magnetic stirrer.

Chemicals:

Chemicals and Reagents	Quantity Taken
Capsicum Pods (paprika)	100.0 g
Petroleum ether (bp- 40°C-60°C)	300.0 ml
Methanolic potassium hydroxide (30%)	100.0 ml
Carbondisulphide	q.s.
Calcium carbonate	q.s.
Benzene	q.s.
Ether	600.0 ml

Principle:

Capsanthin is a rare carotenoid found in paprika. It is a hydroxylated carotenoid present in ester form present in ripe pods of *Capsicum annuum* and *Capsicum frutescens japonicum*. It can be saponified with 30% methanolic potassium hydroxide at room temperature. It can be further purified by column chromatography on suitable adsorbent.

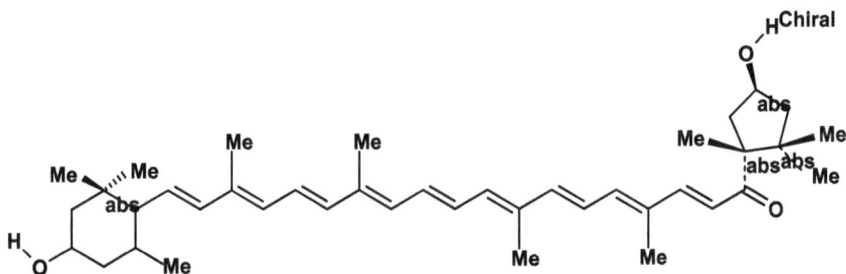

Capsanthin

Procedure:

- Free the paprika pods from their shell and seeds, dry at 35°C-40°C and grind it to fine powder.
- To petroleum ether (boiling point 40°C-60°C, 200.0 ml) add grinded material (100.0 g) and stir for 4.0 hours at room temperature on a magnetic stirrer.
- Filter of the liquid portion with the help of a Buchner funnel and wash the residue with petroleum ether (25.0 ml).
- Dilute the red colored solution of petroleum ether with a 3 fold volume of ether.
- To the above mixture, add methanolic potassium hydroxide (30.0%, 100.0 ml) and stir at room temperature for 8.0 hours to dissolve free phytoxanthins in ether.
- Separate the ethereal layer, wash with water, dry over anhydrous sodium sulphate and concentrate to 20.0 ml under reduced pressure.
- Dilute the ethereal residue with petroleum ether (60.0 ml) and allow it to stand in a cool temperature for 24 hours.
- Filter of the crystals of red capsanthin and recrystallize from small amounts of carbondisulphide. (Extreme care should be taken as carbondisulphide is highly flammable liquid).
- Separte the capsanthin from other carotenoids like zeaxanthin and capsorubin by coloumn chromatography using calcium carbonate and benzene: ether (1:1) as stationary and mobile phases respectively.

Physical Properties:

- Melting point: 176°C.
- Yield: 3.0% $^W/_W$

Identification tests:

- Dissolve a pinch of capsanthin in chloroform (1.0 ml). Add few drops of concentrated Sulphuric acid – A deep blue coloration.
- To capsanthin add antimony trichloride in chloroform – A blue coloration seen.

References:

1. Zechmeister, L.; Von Cholnoky, L. *Ann. Chim.* **1934**, *509*, 269.
2. Karrer, P.; Oswald, A. *Helv. Chim. Acta.* **1935**, *18*, 1303.

## 20

Isolation of Embelin from *Embelia ribes*

Aim: To isolate embelin from *Embelia ribes*.

Apparatus: Column, vacuum distillation assembly.

Chemicals:

Chemicals and Reagents	Quantity Taken
Embelia ribes berries	1500.0 g
Petroleum ether	6000.0 ml
Silica gel	*q.s.*
Benzene	*q.s.*

Principle:

Embelin is chemically 2,5-hydroxy 3-undecyl 1,4-benzoquinone. It is isolated from the berries of the plant *Embelia ribes Burm* belonging to Myrsinaceae Family. It is isolated by using n-hexane by cold extraction method. The purification of the compound is effected by coloumn chromatography using a suitable solvent.

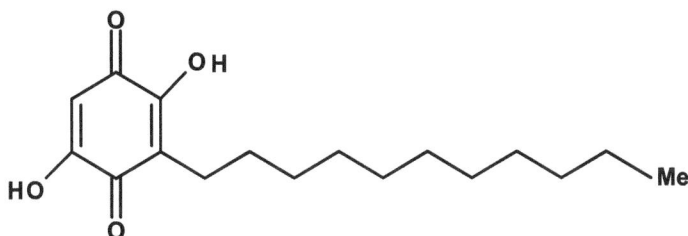

Embelin

Procedure:

- Grind the crude, dry berries of Embelia ribes into coarse powder.
- Extract the powdered berries (1500 g) with n-hexane (6000.0 ml) by cold extraction method.
- Decant the solvent after 72 hours and distill off the solvent over a boiling water bath.
- Concentrate the extract so obtained under vacuum.
- Purify the crude drug so obtained on coloumn chromatography using silica gel and benzene as stationary and mobile phases respectively to give pure orange color powder.

Physical Properties:

- Melting Point: 142°C-143°C.
- Yield: 0.3% $^W/_W$

Category:

- Antifertile, analgesic, anti-inflammatory and anticancer agent.

References:

1. Chitra, M.; Shyamala Devi, C. S.; Viswanathan, S.; Sukumar, E. *Indian J. Pharmacol.***2003**, *35*, 241.

## 21

Counter Current Extraction

1. Counter Current Extraction (CCE)

Discontinuous counter current extraction (distribution) involves a process whereby the 2 liquid phases are allowed to flow counter to each other. It is used for separating materials for isolation or purification purposes. This is specially applicable to the fractionation of organic compounds. Counter current extraction is a process in which contact between phases occurs in a large number, often several hundred, in discrete steps. The method was discovered by L.C. Craig of the Rockefeller institute and is therefore, called *Craig's counter current process*.

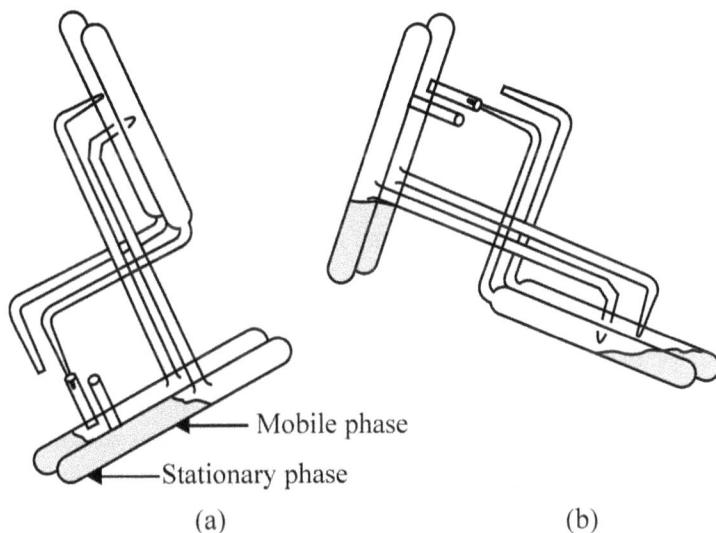

Mobile phase
Stationary phase
(a)   (b)

Fig. 1 Two interlocking glass units for Craig counter-current distrubution. (a) Position during extraction. (b) Position during transfer (Note: By returning the appratus from (b) to (a) the transfer is completed. The mobile phase moves on to the next unit and is replaced by a fresh portion).

The **Craig's method** involves a series of individual extractions, that are carried out automatically in a specially designed apparatus. This consists of a large number (50 or more) of identical interlocking glass extraction units which are mounted in a frame (rack) which is rocked and tilted mechanically to mix and separate the phases during each extraction step. A whole set of extractions is carried out at the same time. At the start of the cycle, all tubes are partly filled with the heavier solvent (*dense phase – extracting solvent*). This is termed the stationary phase as each portion remains in the same unit throughout the procedure.

Tube first unit contains the sample dissolved in the same solvent (solution of the mixture to be separated). A lighter immiscible solvent (mobile phase) is added to the first unit, the phases are mixed and allowed to separate. The upper layer is transferred automatically to the second unit, whilst a fresh portion (not having any sample) is added into the first unit from a reservoir. The extraction and transfer sequence is repeated as many times as there are units. The portions of the lighter solvent (mobile phase) move through the assembly until the initial portion is in the last unit, all units have got portions of both the phases. The first four extractions in sequence for a single solute has been shown in Fig. 2. Here it has been assumed that D = 1 and equal volumes of two phases has been used throughout. As the number of extractions is increased, the solute moves through the system at a rate proportional to the value of D. Many separations are achieved essentially quantitative 95%, 99% or 99.9%. The method has been used for the separation of fatty acids, amino acids, polypeptides and other biological materials. It suffers from the defects that the procedure is lengthy and requires very large volumes of solvents and at present has been replaced by chromatographic techniques.

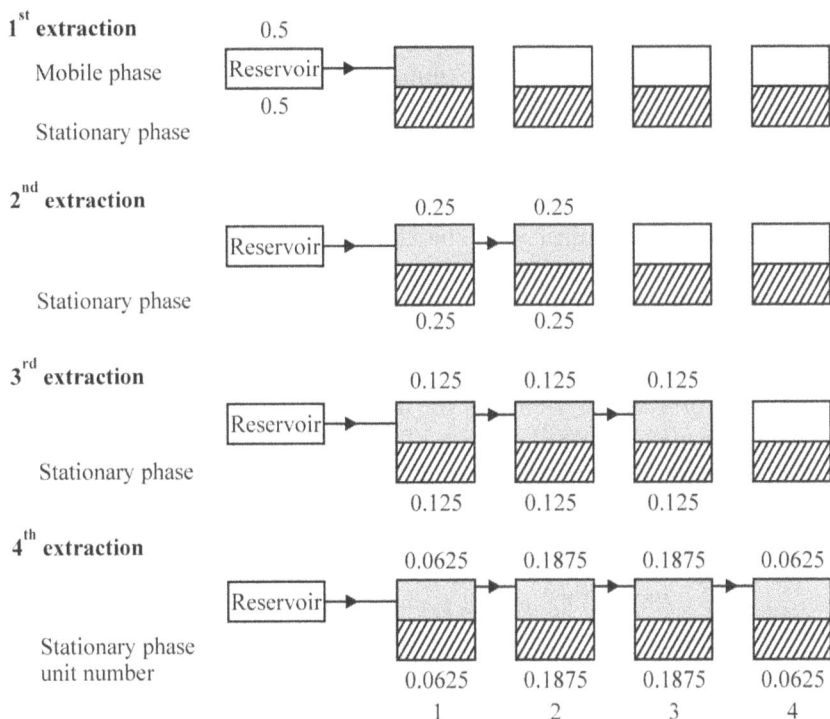

Fig. 2 Extraction scheme for a single solute by Craig counter current distribution. (Figures represent the proportions in each phase for D = 1 and equal volumes. Only the first four extractions are shown).

2. Extraction by Ion-Pair Formation

This is the third largest group of attractants. Many high molecular weight amines and liquid ion exchangers are included here. The mechanism of extraction may be explained as:

$$R_3N_{org} + H^+_{aq} + A^-_{aq} \leftrightarrow R_3NH^+ + A^-_{org} \quad \text{(Extraction)}$$

$$R_3NH + A^-_{org} + B^-_{aq} \leftrightarrow R_3NH^+B^- + A^-_{org} \text{ (Anion extraction)}$$

Another example of an extraction system involving ion pair formation in the organic phase is observed in the use of tetraphenylarsonium chloride to extract MnO_4^-, ReO_4^-, TeO_4^- from aqueous phase into chloroform. The species, an ion pair such as, $[(C_6H_5)_4As^+ \ ReO_4^-]$ passes into the organic phase. Similarly, permanganate forms and ion pair with tetraphenyl arsonium chloride, $[(C_6H_5)_4 As^+ MnO_4^-]$, which makes it organic like and it is extracted into methylene chloride phase. The extraction of uranyl ion UO_2^{2+}, from aqueous nitrate phase into solvents such as ether also involes an ion pair of the type $[UO_2^{2+}. 2NO_3^-]$. The uranyl ion is solvated both by water and ether molecules. The control of temperature, activity, availability of a good diluent and non-formation of emulsions, structure and branching of an amine are some of the factors which provide quantitative separations. The tertiary amines have been found to be the best extractants for the separation of anionic complexes of metals with mineral acids, whereas the para or secondary-amines for the extraction of anionic complexes with organic carboxylic acids. The quarternary ammonium salts (R_4N^+) are relatively good extractant than secondary-amines. Anionic metal complexes like $ZnCl_1^{2-}$, $GaCl_4^-$ and $Co(CN)_6^{3-}$ may be isolated by extraction with tetralkyl ammonium salts, for example:

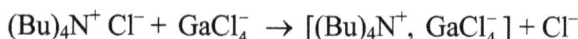

$$(Bu)_4N^+ Cl^- + GaCl_4^- \rightarrow [(Bu)_4N^+, GaCl_4^-] + Cl^-$$

The principal amines used for the purpose are, trioctylamine (TOA), triisoctylamine (TIOA), amberlite, (LA-1), (LA-2) Primene JMT etc.

It is well known that Fe (III) ion can be quantitatively extracted into ether from concentrated HCl medium. The mechanism is not completely understood but the process is useful industrially for the bulk separation of iron prior to the determination of other elements in ferrous alloys. It has been noted that the extractable species is an ion pair of the type $[H_3O^+. Fe(H_2O)2Cl_4^-]$, other equilibria may include:

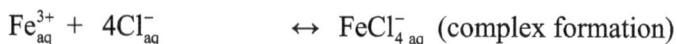

$$Fe_{aq}^{3+} + 4Cl_{aq}^- \quad\quad \leftrightarrow\ FeCl_{4\ aq}^- \text{ (complex formation)}$$

$$R_3N_{org} + H_{aq}^+ + Cl_{aq}^- \quad \leftrightarrow\ R_3NH^+ + Cl_{org}^- \text{ (salt formation)}$$

$$R_3NH_{org}^{+-} Cl^- + FeCl_{4\ aq}^- \leftrightarrow\ R_3NH^+ FeCl_{4\ org}^- + Cl_{aq}^- \text{ (extraction)}$$

The chloro-complex of the iron is coordinated to oxygen atom of the ether (the ether displaces the coordinated water), and this ion associates with a solvent molecule that is coordinated with a proton.

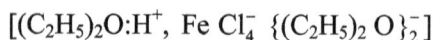

$$[(C_2H_5)_2O:H^+, Fe\ Cl_4^- \{(C_2H_5)_2 O\}_2^-]$$

Thus the chloro-and sulphate-complexes of metal ions can be extracted. One important feature of all these *liquid anion exchangers* is that they are regenerated and are reused just like solid anion exchangers. Certain long chain alkyl ammonium salts, e.g., tricapryl methyl ammonium chloride (aliquant 336 s) and triiso-octylamine hydrochloride (TIOA) are liquids (called liquid anion exchangers), can form extractable ion pairs of aggregates with anionic metal complexes in a similar e.g., in H_2SO_4 medium, uranium is extracted as $2(\text{TIOA-H}^+), UO_2(SO_4^{2-})_2$

The selectivity in such separations can be achieved by controlling various parameters such as, pH for complexation, variation in concentration of an extractant, or the complexing ligand or by selective stripping. The most difficult separations in organic acids such as separation in organic acids such as separation of malonic, tartaric, citric and ascorbic acids are possible to achieve. The *liquid cation exchangers* are also in use for ion pair extractions. Such extractants as bis(2-ethyl hexyl) phosphoric acid (HDEHP) and dinonylsulphonate (DNS) are 2 examples. Here the mechanism of extraction is similar to cation exchange and hence they are called liquid cation exchangers. This mechanism of extraction can be expressed as:

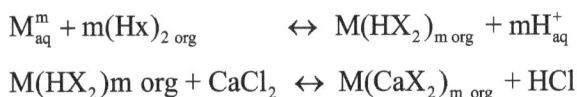

$$M_{aq}^m + m(Hx)_{2\,org} \quad \leftrightarrow \quad M(HX_2)_{m\,org} + mH_{aq}^+$$

$$M(HX_2)m\,org + CaCl_2 \quad \leftrightarrow \quad M(CaX_2)_{m\,org} + HCl$$

The degree of extraction varies with basicity and steric availability of the cation exchanger. The extraction with crown ethers are also an example of the ion pair formation.

3. Solid-Liquid Extraction

Here the separation and isolation of the desired constituent is simply based on the solubility. The extractant (solvent) is chosen such that it has highest solubilising power for the desired substance. Since active constituent present in solid is extracted out by using a liquid solvent it is called a solid-liquid extraction. For extraction, the techniques adopted are (i) maceration, (ii) percolation, (iii) continuous percolation and (iv) Soxhlet extraction. More often extraction of solutes is usually a slow operation and hence a continuous extraction process is preferred. The material is finely ground for most intimate touch with the solvent. The continuous extractors for solids is the Soxhlet apparatus (discontinuous infusion type)

The assembly is quick fit ground glass joint apparatus. The solid is placed in the porous thimble A (made up of hard filter paper) and it is placed into the wide inner tube B. This inner tube has a siphon glass tube and a side arm. It is then fitted into a bolt head flask C, containing the solvent (almost half filled) and to a double surface type reflux the tube E, condensed by the condenser D and drops fall back into the thimble A and in course of time the body of the soxhlet is filled. For solids with low density, the top of the porous thimble A should be above the siphon tube F, otherwise the solid may float out of the thimble and pass down the siphon tube. When the solvent reaches the top of the tube F, it is siphoned to the flask C and thus removes the solute from A to the flask C. The process goes on automatically until complete extraction is achieved. The extracted compound may be separated from its solution by any of the usual methods. It may be noted that the temperature of the solvent in A is quite low as compared to its B.P.; extraction thus takes place at a luke warm temperature and hence is relatively slow. However, if the porous thimble is kept surrounded by the hot vapours of the solvent extraction may be effected by the hot solvent. The substance may also be supported on a sintered glass plate (continuous infusion type. The porous thimble may be substituted by some other convenient receptacles for the solid. Examples of separation by this method are:

(i)   Separation of calcium (Ca), strontium (Sr), barium (Ba) in the form of anhydrous nitrates, only calcium nitrate $(Ca(NO_3)_2)$ is soluble in absolute ethanol or cellosolve where as strontium nitrate $(Sr(NO_3)_2)$ are not.

(ii) Separation of lithium chloride (LiCl) from potassium chloride (KCl) and NaCl, only LiCl is soluble in n-butly alcohol, higher alcohols or in dioxan.

(iii) Separation of potassium chlorate $(KClO_4)$ from lithium (Li), sodium (Na) and magnesium (Mg) perchlorates only $KClO_4$ is soluble in a mixture of equal volumes of n-butanol and ethylacetate.

4. Determination of Lead by the Dithizone Method

One of the most important applications of the solvent extraction technique is in the separation of metal cations in the form of chelate compounds with a variety of organic reagents. The metal cations by this technique (from an aqueous phase into a water immiscible organic phase) can be removed from an interfering matrix, or for selectivity (with the right chemistry) separating one or a group of metals from other. The important chelating agents used are dithizone, oxime, dimethylglyoxime, acetylacetone, thiocyanate, neocuproin and diethyldithiocarbamate etc. The technique is widely used for the spectrophotometric determination of metal ions since the reagents in organic solvents like chloroform ($CHCl_3$) and carbon tetrachloride (CCl_4) form deep colored metal chelates after extraction.

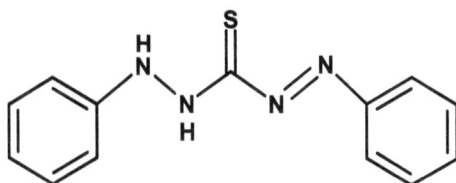

dithizone

The solution in organic solvents of the reagent itself are deep green, where as the metal chelates are intense violet, red, orange, yellow or brown etc., depending upon the metal. Metals, manganese (Mn), iron (Fe), cobalt (Co), nickel (Ni), copper (Cu), zinc (Zn), palladium (Pd), argon (Ar), the determination of lead (Pb) as lead-dithizonate

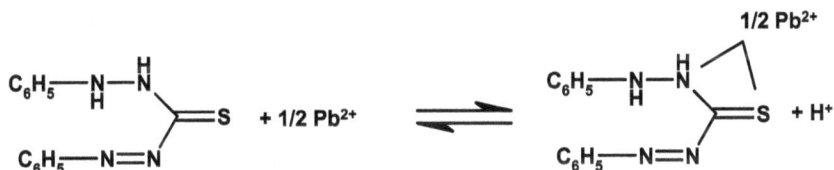

Dithizone acts as a monoprotic acid (pKa =4.7) up to pH -12 the acid proton is that of –SH (thiol) group. Primary metal dithizonates are formed as:

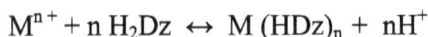

$$M^{n+} + n\,H_2Dz \leftrightarrow M\,(HDz)_n + nH^+$$

Some metals also form a secondary complex (dithizomnates) at a higher pH or with a deficiency of the reagent but these are not of any analytical utility., Dithizone is a violet black solid, insoluble in water, soluble in dilute ammonia solution and also in chloroform ($CHCl_3$) and in carbon tetrachloride (CCl_4) to yield green with it and also it can be made selective by:

(i) **Adjusting the pH of the solution to be extracted:**

Acid solution (0.1-0.5 M) – Argon (Ar) mercury (Hg) copper (Cu) and lead (Pd) may be separated from other metals

Weakly acid solution--- Bismuth (Bi) can be separated from the rest of the metals:

Neutral/faintly alkaline solution- Lead (Pb) and zinc (Zn) can be separated from other metals.

(ii) **Adding a complex forming/or a masking agent-** For example cyanide, thiocyanate, $S_2O_3{}^{2-}$ and EDTA etc.: It is an extremely sensitive reagent and hence its purest form should be used: as the reagent tends to oxidize to diphenylthiocarbadiazone $S=C(N=NC_6H_5)_2$ this does not react with the metals, insoluble in ammonia solution and in organic solvents produces yellow/brown colour. Deionized water or redistilled water should be used. Ammonia solutions should be prepared by passing ammonia gas directly into water. Pyrex vessels should be used and should be rinsed with dill acid before use. Blanks must always be performed. The reagents must be stored in borosilicate glass or polythene bottles.

Reagents:

(i) **Chloroform**: Shake 250 cm^3 of CHCl$_3$ with 25 cm^3 of water containing 1 cm^3 of 10% potassium cyanide (KCN) solution and about 20 drops of 5 M ammonium hydroxide (NH$_4$OH), separate and reject the aqueous layer, wash the CHCl$_3$ layer with water and filter.

(ii) **Dithizone working solution:** Shake 6 cm^3 of a 0.1% solution of dithizone reagent in CHCl$_3$ (which has been filtered and stored in a refrigerator) with 9 cm^3 of water and 1 cm^3 of 5M NH$_4$OH. Separate and reject the lower layer and rotate the aqueous layer in a centrifuge until clear. Prepare this solution afresh on the day of use.

(iii) **Carbamate reagent:** Dissolve 1 g of pure crystalline diethylammonium diethylureidodithiocarbamate in 100 cm^3 of redistilled CHCl$_3$ and store in an amber-coloured bottle. It is only stable for a week.

(iv) **Ammonia-cyanide-sulphite reagent:** Mix 35 cm^3 of concentrated ammonia solution (density 0.88 g/cm^3) and 3 cm^3 if 10% KCN solution (caution) dilute to 100 cm^3 with deionized water and then dissolve 0.15 g of pure sodium sulphite into it.

Procedure I (for samples with a low concentration of Ca, Mg and phosphate). Take the sample solution at room temperature, add 5 cm^3 of 25% ammonium citrate solution together with 10 cm^3 of 10% sodium hexametaphosphate solution. Add few drops of thymol blue indicator and sufficient strong ammonia solution to produce blue-green color i.e., a pH 9.0-9.5. Add 1 cm^3 of regent IV. If much iron is present add 1 cm^3 of 20% ammonium hydroxide hydrochloride (NH$_2$OH.HCl) solution. Transfer it to a 100 cm^3 separatory funnel containing 10 cm^3 of reagent 1 and rinse with a few cm^{3}of water. The volume of the aqueous layer now should be – 50 cm^3 in the separator. Add 0.5 cm of reagent II shake vigorously for a minute and allow to stand. If the lower layer is red, add reagent II until after shaking, a purple, blue or green color is obtained. Run the second separator and wash with 1 or 2cm^3 of reagent I. Shake vigorously the CHCl$_3$ extract. The last CHCl$_3$ extract should be green. If it is not green further extractions with reagent I and II must be made until the green color of the final extract indicates that all the lead has been extracted. Reject the aqueous layer. Add 10 cm^3 of a 1% (1 cm^3 of concentrated HNO$_3$ + 99 cm^3 of deionized water) HNO$_3$ solution in water to the total CHCl$_3$ extract and shake vigorously for a minute. Allow to stand and reject the CHCl$_3$ layer as completely as possible.

Procedure II (for samples with a high concentration of Ca, Mg and phosphate). To the sample add 2-3 drop of methyl-red and make just alkaline by adding strong ammonia solution. Now make the solution just acid with 5 M HCl and add 10 cm^3 in excess. Heat the solution to – 70°C, add 2 cm^3 of fleshly prepared 1.25% sodium metabisulphite solution (filter before use). Transfer it to a separatory funnel at room temperature and adjust the volume 50-75 cm^3 with water in order to bring the concentration of acid to 1 M HCl. Add 10 cm^3 of reagent III by pipette and shake the funnel vigorously for 30 seconds. Allow to stand and run the CHCl$_3$ layer into a 100 cm^3 of regent III. Wash the aqueous layer with reagent I 2-3 times of 1 cm^3 with reagent I 2-3 times without mixing and add this second extract to the main extract. Reject the aqueous layer. To this combined extract add 2 cm^3 of (1:1) sulphuric acid (H$_2$SO$_4$) and evaporate the CHCl$_3$. Add 5 cm^3 of 60% ^{W}/w perchloric acid (HClO$_4$) to the residual solution. Heat until fumes are evolved and the solution becomes clear and colorless. Cool to room temperature, and 10 cm^3 of water and 5 cm^3 of 5 M HCl, boil for a minute, cool, and add 2 cm^3 of 20% ammonium citrate solution,. Now start with "add a few drops of thymol blue indicator"… procedure I.

Determination of Lead: To the HNO_3 layer left in the separator (or your own prepared lead nitrate solution), add 75 cm^3 of the reagent IV (this volumes is taken for 50 ug amount of lead (Pb) if the amount is smaller, adjust accordingly the volume of the reagent IV) adjust the pH of the solution to 9.5 (use a pH meter) by the very careful addition of HCl. Then add 10 cm^3 of reagent 1 and 0.5 cm^3 of the reagent II, shake vigorously for a minute and allow to settle. Run off a little of $CHCl_3$ layer and insert a plug of cotton wool into the dry stem of the separator. After rejecting the first runnings, fill a 1.0 cm absorption cell with $CHCl_3$ layer and carry out blank determination on all the reagents except sample and measure the extinctions at 520 mu of the test and blank solutions against reagent I (all in 1.0 cm cell). Read the ug of Pb equivalent to the observed extinctions from a standard curve and hence know the net amount of Pb in the sample.

*0.0079g of AR lead nitrate are dissolved in deionized water in a 1 cm^3 graduated flask. 10 cm^3 of this solution (containing about 50 ug of Pb) are taken in a 250 cm^3 separator for this determination.

Preparation of standard curve Pipette out into the separator, 0, 1, 2, 3 and 4 cm^3 of saturated lead solution (prepare by dissolving 1.6 g of lead nitrate (AR) in water and adding 10 cm^3 of pure concentrated HNO_3 and making it to $1dm^3$ in a graduated flask with water; this contains 1.0 mg of Pb per cm^3, it is diluted 100 times just before use so that diluted 1 cm^3 now contains 10 mug of Pb). Dilute the contents of each separator to 10 cm^3 with a 1% v/v solution of concentrated HNO_3 in water and now proceed as under determination of lead. Determine the absorbance at 510 mmu with reagent I in the comparison cell (1.0 cm) and draw a curve relating the extinctions to the no. of ug of Pb.

5. **Analysis of Elements in Organic compounds (Oxygen- Flask Method) Oxygen Flask Combustion Technique** as developed by Schoniger (1955) is simple, rapid, elegant, reliable, safe and applicable to the determination of several elements such as halogens (F, Cl, Br, I), sulphur, mercury, phosphorus, arsenic, carbon and boron etc.., in the organic compounds. In pharmacy the method is of special interest as it may be applied directly to the determination of elements in capsules, tablets, lozenges, oily solutions, creams and ointments. In a sealed container, the organic substance is ignited in the presence of oxygen, the organic matter gets destroyed and the inorganic constituents are converted to the forms suitable for determinations by titrimetric or spectrophotometric and/or gravimetric procedures.

A suitable sample size (2–200 mg of the solid compound for semimicro work) is weighed accurately on to a shaped filter paper (of low ash), Fig. 3(b) and folded (or wrapped) in such a way that the tail (or wick) is free. The paper tail serves as the ignition point. The Schoniger apparatus for performing oxidation consists of a heavy walled flask of 300 cm^3-2000 cm^3 capacity fitted with a ground glass stopper [Fig. 3(a)]. Attached to the stopper is 36 mesh platinum gauze basket or carrier (2 cm × 1.5 cm in dimensions) which holds the sample. With ointments, the sample is weighed on to a small square piece of grease proof paper which is folded so as to perfectly enclose the material and then placed in the paper floc to serve as an adsorbent. The methyl cellulose capsules are preferred than gelatin one due to (i) they burn gently; (ii) they give rise to lower blank values; and (iii) no acidic products are formed on combustion.

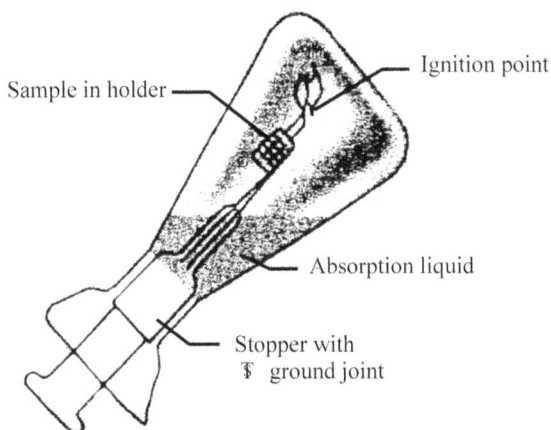

Fig. 3(a) Schoniger combustion apparatus.

The flask contains a few cm^3 of an absorbing solution (e.g. aqueous sodium hydroxide (NaOH) / sodium carbonate (Na$_2$CO$_3$), but would vary in accordance with inorganic element to be determined) and the air in the flask is displaced by oxygen. The tail of the paper is ignited and the flask is then sealed immediately using the stopper with the platinum gauze basket. Or, better it may be ignited by a remote electrical control or, an infrared lamp. The flask is inverted to prevent the escape of the volatile oxidation products and also to prevent incompletely burned material from falling out of the gauze into the liquid. Hold the stopper firm in its place during combustion of the sample. The combustion is rapid (being catalyzed by the platinum (Pt) gauze surrounding the sample) and generally complete in 5-10 seconds. After allowing to stand for a few minutes until any combustion gases have disappeared, the flask is shaken vigorously for about 5 minutes to ensure complete absorption. In some cases a small amount of carbon may deposit on the wall of the flask during combustion but this does not appear to have any effect on the result.

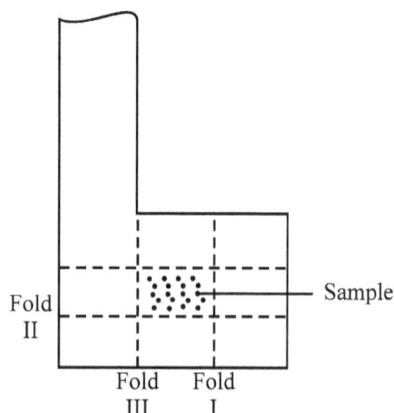

Fig. 3(b) Paper shape for placing sample.

Fig. 3(c) Stopper with Pt-wire and gauze used in
oxygen flask combustion method.

After cooling, the flask is shaken and a few cm³ of water is placed in the collar of the flask and the stopper is disassembled. The water washes and rinses down the neck. The stopper, pt-wire and pt gauze and walls of the flask are rinsed with water. The resulting solution is then analyzed by some suitable method applicable to the particular element being determined.

The size of the flask (500 cm³) is suitable for the semi micro determination of most of the organic compounds (sample size 50 mg). However, low content formulations are determined using a bigger sample in a 1 dm³ or even in a 2 dm³ flask. A sample of 50 mg is burnt in a 500 cm³ flask without the formation of carbon. Samples of the size (e.g. thyroid) weighing 0.6 g-0.7 g have been burnt in a 2 dm³ flask.

Determination of Individual Elements

1. **Chlorine containing substances:** The sample taken is usually 50mg in a 500 cm³ flask. The absorbing liquid consists of 20 cm³ of water and 1 cm³ of hydrogen peroxide (H_2O_2) (100vol.) Generally 3 methods are used : (i) a potentiometric titration with silver nitrate ($AgNO_3$); (ii) a titration with mercuric nitrate ($Hg(NO_3)_2$) using diphenyl carbazone as indicators in an ethanolic medium; and (iii) a Volhard's procedure.

    (i) **Potentiometric titration with $AgNO_3$:** Add a drop of 0.1 % dilution of methyl red in 95% CrH_5OH to the resulting liquid in combustion flask. Neutralize by addition of 1N NaOH solution and then add in excess, 0.5 cm³ or more. Boil gently for 5-10 minutes (using a reflux condenser to avoid loss of water) to eliminate the peroxide. Cool the flask and transfer quantitatively its contents to a 250 cm³ beaker. Add a drop of methyl red indicator and then 1 N HNO_3 until the indicator changes to red in color. Next add 2 cm³ of acetate buffer (50 cm³ of glacial acetic acid = 50 g of sodium acetate ($NaOOCCH_3$) per dm³). Immerse the beaker in an ice-bath so that the contents are cooled to 8°C-10°C. Now titrate potentiometrically with 0.02 N $Hg(NO_3)_2$ solution using either calomel /(KNO_3) or Silver/silver chloride (AgCl) electrodes.

    (ii) **Mercuric nitrate titration**: Here use about 3 cm³ of 0.1 N NaOH more in the absorption liquid. Add 2 drops of 0.1% solution of the bromophenol blue in ethyl alcohol (C_2H_5OH) and next add 0.1 N HNO_3 to the yellow green color of the indicator and 0.5 cm³ of the acid more. If the compound contains sulphur, add a little excess of 0.1% solution of diphenyl carbazone in (C_2H_5OH) and titrate with a 0.002 N $Hg(NO_3)_2$ solution from a 5 cm³ micro burette to the first appearance of a pink shade. Carry out a blank determination. Calculate as :

$$1 \text{ cm}^3 \text{ of } 0.02 \text{ N Hg(NO}_3)_2 = 0.0007092 \text{ g Cl}$$

Preparation and standardization of 0.02 N Hg(NO₃)₂ solution: Dossolve 6.86 of $Hg(NO_3)_2$ in 75 cm^3 of 0.1 N HNO_3 and dilute to about 1500 cm^3 with water. Add 200 cm^3 of 95% C_2H_5OH dilute to 2 dm^3 with water and if necessary, filter. The solution is usable only for a week.

0.02 N NaCl solution: Weigh 1.1690 g of pure (AR) NaCl, previously dried at 250°C for 2 hours and dissolve in a 1 dm^3 graduated flask with

Titration: Take a 250 cm^3 conical flask and place 5 cm^3 of 0.02 N NaCl solution into it, add 15 cm^3 of water, 2 drops of bromophenol blue (0. 1% solution in C_2H_5OH) indicator and 0.1 N HNO_3 to the yellow – green colour of the indicator and 0.5 2 of the acid more. Next add 100 cm^3 of 95% ethanol and 15 drops of 0.1% solution of diphenylcarbazone in ethanol and titrate with the $Hg(NO_3)_2$ solution (from a 5 cm^3 microburette) to the first appearance of a pink shade. Determine the blank and subtract this, from the previous titre value. Calculate the normality of the $Hg(NO_3)_2$ solution.

(i)   **Volhard's Method:** Take absorbing liquid as 20 cm^3 of 1 N NaOH solution. To the total solution and washings in a beaker add 2.5 cm^3 of 5 N HNO_3 and 10 cm^3 of 0.1 $AgNO_3$ solution in excess and 1-2 cm^3 of nitrobenzene, add titrate with 0.05 M Ammonium thio cyanate (NH_4 SCN) solution using 3-4 drops of ferric alum solution (dissolve 0.2g of alum in 50 cm^3 water + 6 cm^3 of 5 N HNO_3 and then diluting to 100 cm^3 with water) as indicator. Carry out a blank titration. The difference between the two titrations indicate the amount of $AgNO_3$ consumed for the sample.

$$1 \text{ cm}^3 \text{ of } 0.1 \text{ N (Ag NO}_3) = 0.001773 \text{ g of (Cl)}$$

2. **Bromine containing substances:** Here the absorbing liquid consists of 20 cm^3 of a 30% AR NaCl solution, 10 cm^3 of sodium hypochlorite solution and 5 cm^3 of 20% buffer solution of sodium dihydrogen phosphate. Heat the solution to boiling, add 5 cm^3 of 50% sodium formate solution and again boil for a few minutes. Draw air through the flask to remove chlorine vapors. Cool and dilute to 100 cm^3 with water. Add 1g potassium iodide (KI), 25 cm^3

of 6 N H_2SO_4 and 1 drop of 0.5 N ammonium molybdate, Titrate immediately with 0.02 N sodium thiosulphate solution using starch as indicator and carry out also a blank determination.

Bromine may also be determined by the $Hg(NO_3)_2$ titration method as under chlorine containing substances. In such a case the absorbing liquid consists of 20 cm³ of water, 1 cm³ of H_2O_2 (100 vol) and 3 cm³ of N/10 NaOH. Calculate as:

$$1 \text{ cm}^3 \text{ of } 0.02 \text{ N } Hg(NO_3)_2 = 1.5984 \text{ mg of Bromine}$$

3. **Iodine containing substances:** A mixture of 10 cm³ of water and 2 cm³ of 1 N NaOH is used as the absorbing liquid. Add an excess of acetic bromine** solution (5 cm³-10 cm³) and allow to stand for about 2 minutes: the bromine in potassium acetate ($KOOCCH_3$) – acetic acid buffer solution completes the oxidation of iodine to iodide. Add formic acid HCOOH (0.5 cm³-1 cm³) to eliminate excess of bromine (reduction to hydrogen bromide (HBr)) and remove any vapours of bromine (Br_2) from the flask by passing air. Add 1g KI and immediately titrate with 0.02 M sodium thiosulphate ($Na_2S_2O_3$)-5H_2O solution using starch as indicator

4. **Fluorine-containing substances:** The combustion is carried out in a soda-glass flask (totally free from boron). The best results are obtained by using a flask made up of fused silica glass. The absorbing liquid is simply 20 cm³ of water. After combustion of the sample add enough water to make a volume of 50 cm³. Take 2 cm³ of this liquid expected to contains about 250 g of fluoride (from the combustion flask) add 20 cm³ of water, 10 cm³ of alizarin fluorine blue solution and 3 cm of buffer (2 % sodium acetate ($NaOOCCH_3$) + 6% $^V/_V$ glacial acetic acid) together with 10 cm³ of $Ce(NO_3)_3$ and enough water to make 50 cm³. Allow to stand in the dark for an hour. Then measure the extinction in 1cm cell at 610 nm. Prepare a blank similarly using 2 cm³ of water in place of the sample. Calculate the amount of fluorine from a calibration curve prepared by using 22 mg/2 dm³ sodium fluoride (NaF) solution.

5. **Sulphur-containing substances:** The absorption liquid consists of 15 cm of water and 1 cm³ of H_2O_2 (20 volume) 2 methods of estimation are as:

    (i) *Substances which do not produce acidic/basic substances except H_2SO_4 on combustion:* Boil the solution for 10 minutes to destroy peroxide and then cool. Titrate with a standard NaOH solution (0.05 N or 0.02 N depending upon the quantity of the acid produced) and use screened methylred as indicator.

(ii) *For all Other sulphur-containing substances:* After combustion wash down the stopper, platinum wire, pt-gauge and inner surface of the flask with about 60 cm^3 of industrial methylated spirit. Add 2 drops of a 0.2% aqueous solution of the sodium salt of 1-(o-arsono-phenyl-azo)-2-napthol-3,6-disulphonic acid (Thoron) and 2 drops of a 0.0125% aqueous solution of methylene blue. Titrate with 0.02 M or 0.01 M barium per chlorate $Ba(ClO_4)_2$ solution until the yellow color changes to pale pink. The titration is carried out in good natural light and the contents may preferably be vigorously stirred by means of a magnetic stirrer.

The Organophosphorus compounds, which are subjected to combustion, give mainly orthophosphate and the latter can be absorbed in either H_2SO_4 or HNO_3 and the phosphorus present is readily determined spectrophotometrically either by the *molybdenum blue method or as the phosphovanadomolybdate.* Procedures have also been extended for determining of metallic constituents. Thus Hg compounds from organomercury is absorbed in HNO_3 and titrated with sodium diethydithiocarbamate; Hg is also determined by an EDTA titration solution, also Zn from organo zinc compounds is absorbed in HCl and determined by an EDTA titration. The flask combustion analysis also includes the determination of water in animal tissue and blood samples prior to measurement if their C^{14} and tritium contents.

At present the organo-iodine compounds assayed by this technique are, nitroxynil, propyliodine, liothyronine sodium, thyroxine sodium, propyliodine, oily suspension and lophendylate injection etc. Similarly organo-chlorine compounds like cloxacillin sodium, niclosamide etc.., are assayed for chlorine through this procedure. Flourine in fluocinolone acentonide (for F) is assayed similarly. This technique also used to confirm the presence of chlorine (Cl_2) qualitatively in diloxamide furoate and chlorthalidone etc.

5. **Estimation of Nitrogen (Kjeldahl Method)**

Theory: The nitrogen – containing organic compound is digested with concentrated H_2SO_4 in the presence of a catalyst where by the nitrogen of the sample is converted into ammonium sulphate $(NH_4)_2SO_4$. Afterward an excess of standard NaOH solution is added and the resulting ammonia is steam distilled and absorbed in excess of the standard H_2SO_4 solution. The remaining acid is back – titrated against standard NaOH solution.

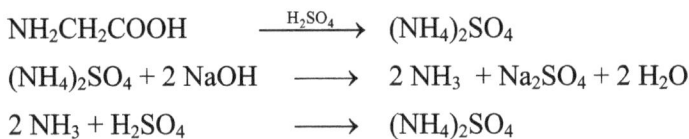

$$NH_2CH_2COOH \xrightarrow{\;H_2SO_4\;} (NH_4)_2SO_4$$

$$(NH_4)_2SO_4 + 2\ NaOH \longrightarrow 2\ NH_3 + Na_2SO_4 + 2\ H_2O$$

$$2\ NH_3 + H_2SO_4 \longrightarrow (NH_4)_2SO_4$$

Chemicals required:

Sulphuric acid solution (0.1 N)

Sodium hydroxide solution (0.1 N, 4 g/L)

Catalyst: [a mixture of K_2SO_4 (20 g), selenium powder (1 g) and $CuSO_4.5H_2O$(1 g)]

Procedure: Weigh accurately 0.1-0.2 g of the nitrogenous organic compound (benzamide, diphenylamine, acetanilide glycine, etc) in a clean 50 cm^3 Kjeldahl flask. Add 1.0 g of the catalyst and 5cm^3 of concentrated H_2SO_4. The K_2SO_4 serves to elevate the boiling point of H_2SO_4 while selenium and copper (II) sulphate act as the catalysts.

Stopper the flask loosely and digest on a sand bath as shown in Fig. 4(a) in a slightly inclined position for 2-3 hours in a hood. This is the minimum time required for the decomposition and digestion. The duration may be more for some other compounds, for example, for polymeric nitrogenous organic compounds. This duration may be from 10-12 hours. After the digestion is completed, cool the flask. In the meantime set up the distillation apparatus as depicted in Fig. 4(b). Transfer the mixture from the Kjeldahl flask into the chamber C. Pass steam into the outer chamber B by turning the stopcock A. Introduce carefully 40 cm^3 of 50% NaOH solution throughout the funnel E. Take 50 cm^3 of standard H_2SO_4 in a 250 cm^3 conical flask G. Continue to pass the steam for about an hour to ensure that all the ammonia evolved has been absorbed by the acid. Turn off the stopcock A to cut off the acid. Turn off the stopcock A to cut off the flow of steam.

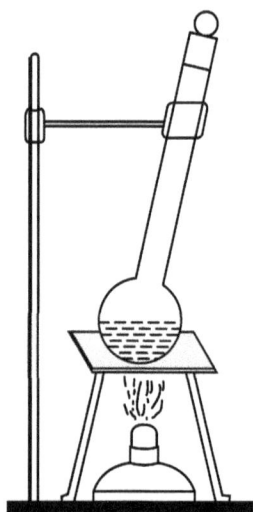

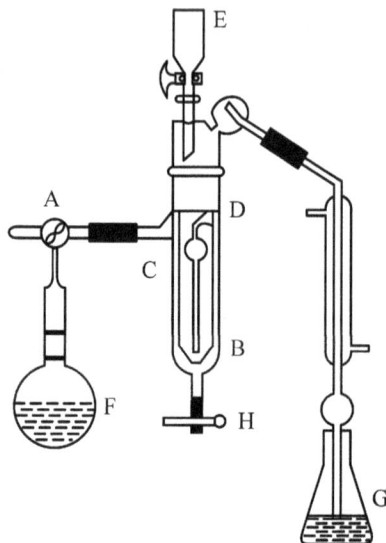

(a)   (b)

Fig. 4 (a)   **Fig. 4 (b)**

The excess of the acid in the conical flask may be titrated with a standard NaOH solution using phenolphthalein as an indicator.

Calculations:

Weight of the nitrogenous compound $= W$ g

Volume of 0.1 N H_2SO_4 taken $= 50$ cm^3

Volume of 0.1 N NaOH solution required to back-titrate the excess acid $= V$ cm^3

Volume of 0.1 N H_2SO_4 used up for reactions with the ammonia evolved $= (50 - V)$ cm^3

$(50 - V)$ cm^3 0.1 N $H_2SO_4 = (50 - V)$ cm^3 of 0.1 N (NH_3)

1000 cm^3 of 1 N $NH_3 = 17$ g of ammonia $= 14$ g of nitrogen

1000 cm^3 of 0.1 N NH_3 corresponds to 1.4 g of nitrogen $(50 - V)$ cm^3 of 0.1 N of NH_3 corresponds to $(50 - V) \times 1.4/1000$ g of nitrogen

Percentage of nitrogen $= (50 - V) \times \dfrac{1.4}{1000} \times \dfrac{100}{W}$

Estimation of sulphur (Messenger's Method)

Theory: Sulphur in an organic compounds estimated by the Messenger's method. In this method sulphur of the organic compound is converted to sulphuric acid by heating with an alkaline potassium permanganate solution. Sulphuric acid is then precipitated with the barium chloride solution as barium sulphate and is thus estimated gravimetrically. The reactions that take place (if for example thiourea is taken) are:

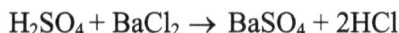

$$H_2NCSNH_2 + 8KMnO_4 + 8KOH \rightarrow H_2SO_4 + H_2NCONH_2$$
$$+ 8K_2MnO_4 + 3H_2O$$

$$H_2SO_4 + BaCl_2 \rightarrow BaSO_4 + 2HCl$$

Chemicals Required:

Potassium permanganate (5 g - 6 g)

Barium chloride (5%) solution

Potasium hydroxide (10%) solution

Procedure: Place 0.2 g - 0.25 g sample of thiourea in a 250 cm^3 round bottomed flask fitted with an air condenser. To this add 50 cm^3 of distilled water and 5 cm^3 of 10% NaOH solution. Shake the contents thoroughly and subsequently add 4-4.5 g of potassium permanganate. Add a few small pieces of porcelain chips, fit in the condenser and reflux the contents for 4-5 hours on a sand-bath. If the color of the KMnO$_4$ disappears during heating, add more of it. After this duration, cool the contents to room temperature and add concentrated HCl (10 cm^3-20 cm^3) drop wise to the flask until the purple color of unreacted KMnO$_4$ is decolorized. Heat the flask again for 10-15 minutes and filter the contents into a 250 cm^3 beaker. Wash the flask thoroughly and transfer the washings on the filter paper. Concentrate the solution to about 50 cm^3. Add barium chloride solution slowly and with constant stirring in order to precipitate sulphate (SO_4^{2-}) sulphuric acid produced, as barium sulphate. Allow the beaker to stand undisturbed for at least 4 hours (preferably overnight) and filter the precipitate on a Whatman filter paper (of fine porosity). Wash the precipitate with distilled water till whole of the Cl$^-$ ions get removed (i.e., no precipitate is formed with AgNO$_3$ solution). Fold the filter paper, place it in an alreadyconstant weighed silica crucible and heat on a non-luminous Bunsen burner flame to make the ash white. Cool the crucible in a desiccator and weigh to a constant weight.

Calculations : Weight of the organic compound sample = W g

Weight of the barium sulphate precipitate = w g

According to the reaction 233.4g of $BaSO_4$ = 32 g of S

Therefore, w g of $BaSO_4$ = (32/233.4) × w g of S

If the weight (W) of the sample is known, then the percentage of

$$sulphur = \frac{32}{233.4} \times W \times \frac{100}{W}$$

7. Determination of Moisture and Water (Karl Fischer Reagent):

The Karl Fischer titration method is used for the determination of water in various types of solids and organic liquids. The titration method is based upon oxidation/ reduction which is specific for water. The reagent was proposed by Karl Fischer in 1935 and prepared it by the action of sulphur dioxide (SO_2) on a solution of iodine (I_2) in a mixture of anhydrous pyridine and anhydrous methanol (solvents which are neither too acidic nor too basic-aprotic solvents) as the solvents. A large excess of pyridine was used to complex the I_2 and SO_2. Water reacts with this reagent in two steps. In the first step, I_2 and SO_2 in the presence of pyridine and water react to form pyridinium sulphite and pyridinium iodide. In the second step pyridinium sulphite reacts with methanol to form pyridinium methyl sulphate.

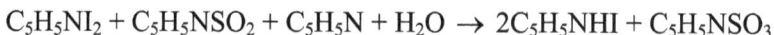

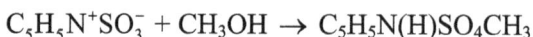

$$C_5H_5NI_2 + C_5H_5NSO_2 + C_5H_5N + H_2O \rightarrow 2C_5H_5NHI + C_5H_5NSO_3$$

$$C_5H_5N^+SO_3^- + CH_3OH \rightarrow C_5H_5N(H)SO_4CH_3$$

The pyridinium sulphite can also consume water.

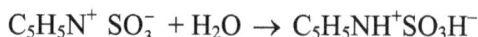

$$C_5H_5N^+ SO_3^- + H_2O \rightarrow C_5H_5NH^+SO_3H^-$$

However, this last reaction is prevented by using a large excess of CH_3OH and this it is not specific for water. It may be noted that one molecule of I_2 disappears for each molecule of water present.

The reagent prepared with an excess of methanol was unstable and needed frequent standardizations. Later on by using 2–methoxyethanol, 2-chloroethanol or trifluoroethanol in place of methanol, the stability of the reagent was increased and it was increased and it was described as "methanol free". However, it is essential to have an alcohol in this reagent to esterify the sulphur dioxide (SO_2). A base is necessary to neutralize the acids produced in this reaction.

$$I_2 + SO_2 + 2H_2O \rightarrow 2HI + H_2SO_4$$

The pyridine is a too weak base to completely neutralize such acids and hence gives sluggish end points. Now the base used is imidazole and the titration is done in the pH range 5-7. Such a karl Fisher reagent (KFR) is described as **"Pyridine free"**.

The method is limited to those cases where the test compounds does not react with any component of the reagent or with HI or water. However, the following compounds interfere in Karl Fischer titration:

(i)   Oxidants like CrO_4^{-2}, $Cr_2O_7^{-2}$, Cu(II) and Fe(III) salts, higher oxides and peroxides

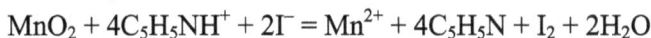

$$MnO_2 + 4C_5H_5NH^+ + 2I^- = Mn^{2+} + 4C_5H_5N + I_2 + 2H_2O$$

(ii) Reductants Like $S_2O_3^{2-}$, Sn(II) salts and sulphides.

(iii) Substances which form water with the components of the Karl Fischer reagent. Examples are basic oxides and salts of weak oxyacids.

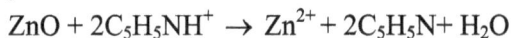

$$ZnO + 2C_5H_5NH^+ \rightarrow Zn^{2+} + 2C_5H_5N + H_2O$$
$$NaHCO_3 + C_5H_5NH^+ \rightarrow Na^+ + H_2O + CO_2 + C_5H_5N$$

(iv) Aldehydes form a bisulphate addition compound.

$$RCHO + SO_2 + H_2O + C_5H_5N \rightarrow RCH(OH) SO_3 HNC_5H_5N$$

(v)   Ketones react with methanol to produce ketal and water.

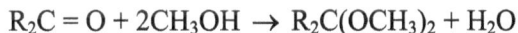

$$R_2C = O + 2CH_3OH \rightarrow R_2C(OCH_3)_2 + H_2O$$

The pyridine free reagents are available commercially for both titrimetric and colorometric Karl Fischer procedures. The reaction is now considered to proceed as:

Solvolysis $2ROH + SO_2$ $\leftrightarrow$ $RSO_3^- + ROH_2^+$

Buffering $B + RSO_3^- + ROH_2^+$ $\leftrightarrow$ $BH^+ SO_3R^- + ROH$

Redox $B.I_2 + BH^+ SO_3R^- + B + HO_2 \leftrightarrow BH^+SO_4R^- + 2BH^+I^-$

However, the Stoichiometry is again one mole of I_2 used for each mole of H_2O, present in the sample.

The KFR is deep reddish brown in color when fresh. The used reagent possesses a pale straw-yellow color which itself serves as an indicator in visual titrimetry. The reagent is unstable and undergoes auto-oxidation with time. The decomposed reagent has a brownish dark color. Hence it is placed in ground glass stoppered bottles or in reservoirs guarded against atmospheric moisture.

Preparation of the Karl Fischer reagent (KFR): The components of the reagent must be pure and free from water. Ethylene glycolmonomethyl ether (methylcellosolve) normally has a water content of < 0.1% and can be used without further purification. If the water content is known to be more than this, it should be purified by distillation.

Place 400 cm^3 of dehydrated (CH_3OH) (water content < 0.03%) into a glass stoppered flask of 750 cm^3 capacity and then mix 80 g of the dehydrated pyridine. Immerse the flask in a freezing mixture and pass a slow current of dry SO_2 gas into this cold solution with continuous stirring until the increase in weight is 20 g. Care is taken at all times to prevent access of moisture. Add 45 g of I_2 (resublimed) and shake well to dissolve. Allow the mixture to stand for 2 hours before use. Store the bulk of the reagent in a glass stoppered container, completely protected from light. It has a water equivalent of 3.5 mg/cm^3 approximately.

Apparatus: Various forms of apparatus for use in indirect titration are available commercially. The apparatus described here is suitable for samples having water contents of about 0.1%. Fig. 5 represents the main features of the apparatus. The KFR and the standard solution of water in CH_3OH are contained in 2 separate bottles and these solutions are pumped into the respective burettes by means of hand bellows. The entry of moisture is prevented by a suitable arrangement of drying tubes. The reaction vessel should be at least of working capacity of 60 cm^3, the side tube being provided for introducing the liquid samples. The burettes are of 10 cm^3 capacity, graduated in 0.02 cm^3 and are closed by the drying tubes. The reaction vessel is fitted with an air tight rubber closure through which pass 2 platinum electrodes, jets of the burettes, and a vent tube filled with a desiccant. A glass tube for passing dry nitrogen (N_2) gas is also introduced into it. Stirring is done magnetically. The galvanometer (100 µA full scale deflection) should have a resistance of less than 100 Ω. Other loose apparatus required include 10 cm^3 and 20 cm^3 pipettes, a weighing pipette, all should be fitted with hypodermic needles (1.0 mm bore), graduated flask (100 cm^3 with latex vaccine caps, 2 or 3 activated silica gel drying bottles and a final wash bottle having KFR for drying (N_2) gas supply.

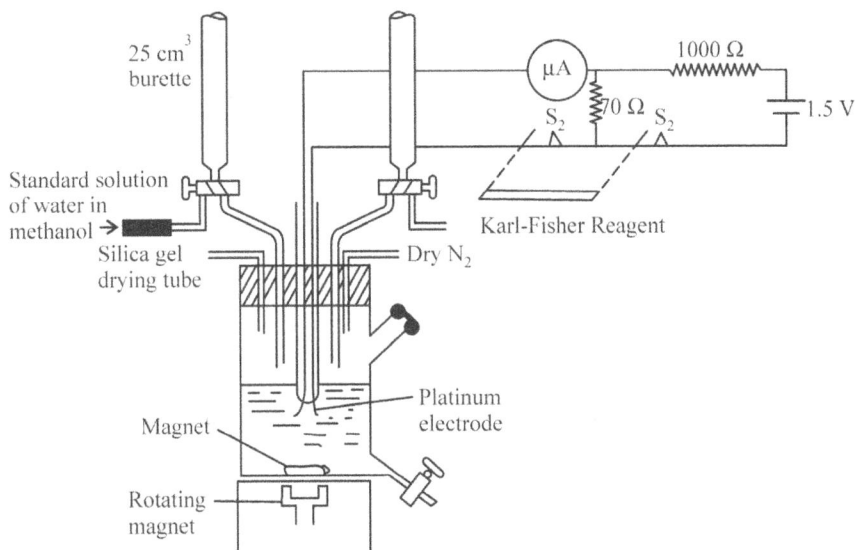

Fig. 5 Preparation of Karl-Fisher reagent.

Precautions:

(i) Prevent atmospheric moisture from entering the apparatus and this is achieved by using suitable guard tubes filled with fresh activated silica gel.

(ii) Use the colorless silicone grease (does not react with KFR) to lubricate the stop-cocks.

Detection of end point: In titration with KFR, only the cathode is polarized, the anode is depolarised by iodine (I^-) ions. At the end point free iodide (I_2) depolarizes the cathode and hence the current rises rapidly, giving a deflection which persists for at least 30 seconds.

When titrating with the water-methanol (H_2O-CH_3OH) reagent, both electrodes are depolarized and hence the current is large. At the end point (all the I_2 is reduced) the iodide ions polarize that cathode and hence the current reduces almost to zero. The zero of the meter scale is reached only when H_2O-CH_3OH reagent is added in excess. An end pint in a KF titration can be noted visually based on the brown color of the excess reagent. More often end points are observed electro analytically. Several instrument manufactures supply automatic or semiautomatic instruments for KF titration and all these are based on electrometric end point detection.

Standardization: KFR is not stable and is absolutely protected against light and moisture. It must be standardized immediately before use by titration against a known amount of water, either as a standard water-methanol reagent or as solid crystalline sodium titrate dehydrate. The sodium compound is almost insoluble in methanol; however it transfers its water of hydration to the solvent. The water content can be determined by drying at $150°C$ to a constant weight. All glassware must be carefully dried before use. It should also be attempted to minimize contact between the atmosphere and the solution during the titration.

With water-methanol reagent

1. Take a dry glass Stoppered 1 dm^3 graduated flask and about 900 cm^3 of anhydrous CH_3OH in it (0.1% of H_2O) and place it in a thermostatically controlled water bath at $25°C$ together with a small flask having about 200cm^3 of the same CH_3OH. Weigh out accurately about 15 g of distilled water into the 1 dm^3 flask and after the contents are at the temperature of the bath, adjust the volume to the mark with CH_3OH from the smaller flask.

2. Transfer (allow to run) sufficient KFR (from its vessel) to the reaction vessel to cover the electrodes. Pass a current of dry N_2 gas through the apparatus. Switch on magnetic stirrer and wait as such for 15 minutes, so that even traces of moisture get remove. Close the switch S_1 and S_2 and run in sufficient water-methanol reagent to decrease the meter reading to about 10 μA and wait for 15 minutes more as such. The galvanometer reading should remain constant during all this period thereby indicating that the apparatus is dry and free of air leaks. Add more H_2O-CH_3OH reagent so that the meter reading becomes about 2 μA. At this stage the reaction vessel may be drained, if necessary, run in exactly 10 cm^3 of KFR and titrate with H_2O-CH_3OH reagent until again the meter shows a reading of 2 μA. Repeat the titration to agree within 0.1cm^3 open the switch S_1/ S_2. Calculate the strength of H_2O-CH_3OH reagent vs KFR as:

$$1\ cm^3\ H_2O - CH_3OH\ reagent = M\ cm^3\ KFR$$

3. Most of the liquid in the reaction vessel is drained off. Transfer into it 10 cm^3 of anhydrous methanol by means of a dry pipette. Pass dry nitrogen gas through the apparatus continuously. Start the magnetic stirrer and close the switch S_1/S_2. Add the KFR until a definite brown color exist for at least 20 seconds, when the meter should read 80 uA (depends upon the size and distance a part of the electrodes). Note down the volume of KFR added, let it be v_1 cm^3. Titrate with the H_2O-CH_3OH reagent until the meter shows small positive reading of

about 2 μA and note down this volume, let it be v_2 cm^3. Repeat the titrations until they agree within 0.1 cm^3. Repeat the titration by taking 10 cm^3 of the standard solution of water in methanol. Note down the volume of KFR used (let it be v_3 cm^3) and H_2O-CH_3OH reagent (let it be v_4 cm^3) added. Open the switch S_1/ S_2.

Calculate the water equivalent (F), in mg per cm^3 of the KFR as:

$$F = \frac{100\ w}{(v3 - v1) + M\ (v2 - v4)}$$

The water content of the water-methanol reagent is = MF mg/ cm^3

This value may be use for the further re-standardization of KFR.

Standardization with sodium tartarate dehydrate: Dry about 1 g of sodium tartarate dehydrate at 130°C to a constant weight overnight; the water content of the dried material is 15.65%w/w.

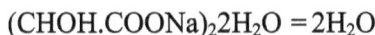

$$(CHOH.COONa)_2 2H_2O = 2H_2O$$

or,   230.1 g $(CHOH.COONa)_2 2H_2O = 2H_2O = 36.0304$ g H_2O

or,   1 g $(CHOH.COONa)_2 2H_2O = 0.1565$ g H_2O

Introduce an accurately measured volume of KFR (V3 cm^3) into the reaction vessel (clean and dry); the volume should be such as to cover the electrodes. Titrate electrometrically with the solution of water in CH_3OH. Repeat until 2 titrations are in agreement within 0.1 cm^3. Let this final volume required be v_1 cm^3.

Weigh about 0.2 g of the dried sodium tartarate dehydrate and introduce it into the reaction vessel, add v_3 cm^3 of the KFR and again titrate with the solution of water in methanol. Repeat the procedure until the 2 successive titrations are in agreement within 0.1 cm^3. Let the final volume required be V_2 cm^3.

If F is the mg of water in 1 cm^3 of the solution of water in CH_3OH and w (wt in g) of sodium tart rate dehydrate used. Then:

$$F = \frac{15.65 \times 1000 \times w}{(v1 - v2) \times 100}$$

8. Experimental determination of Moisture (Aquametry):

Introduce a suitable volume (x cm^3) of the KFR into the reaction vessel sufficient to cover the electrodes. Allow to pass a current of dry N_2 gas into the assembly. Now titrate electrometrically with a standard solution of water ion CH_3OH. Repeat the procedure and note that the 2 successive titrations are in agreement. Let this final volume required be a cm^3. Introduce, now, an accurately weighed or measured quantity of the sample into the reaction vessel and add the same (x cm^3 as above) of the KFR and again titrate with a standard solution of water in methanol. Let this time the volume required be b cm^3.

$$\% \text{ of water (in sample)} = \frac{(a - b) \times F \times 100}{100 \times \text{wt (or vol.) of sample taken}}$$

Applications: The KFR method has been applied to the determination of water/ moisture in vary many types of samples. It is used to determine water in many types of samples. It is used to determine water in may organic acids, alcohols, ester, ethers, anhydrides and halides, hydrated salts of organic acids and inorganic salt hydrates (soluble in CH_3OH) are readily determined by direct titration. The direct and rapid titration is feasible, if the sample can be dissolved completely in CH_3OH. It has been used to determine water in antibiotics (erythromycin, gentamicin sulphate) in steroids (hydrocortisone sodium phosphate) and sodium nitroprusside, ca-acetate demethyl phthalate, ephedrine, fructose, benzyl penicillin in sodium, isopropyl alcohol, potassium citrate and isopenaline sulphate, saccharin sodium etc.

Sometimes in case of samples, which are partially dissolved in KFR and given incomplete recovery of water, are first refluxed for quite sometimes with anhydrous CH_3OH or other organic solvents. The resulting solution is then titrated directly with that reagent. In some of such cases, satisfactory results are obtained by adding excess of the KFR reagent and back titrating with a standard solution n of water in CH_3OH after allowing a suitable reaction time.

Part B

Solvent Purification Techniques

Introduction

Commercially available grades of organic solvents are of sufficient purity to use in many reactions as far as small quantities of water or any other impurities present in them does not affect the reaction process and induce unlikely side reactions. However if it is affecting the reaction course, it is necessary to purify the solvents by removing water and other impurities. Though analytical grades of many of these solvents are available, it is advisable to purify ourselves as it will be a cost effective.

To remove water from the solvents/solutions of organic compounds, various solid drying agents are utilized. The characteristic features of an ideal drying agent otherwise called as dehydrating agent, desiccant are as follows.

- It should not react chemically with organic solvents.
- It should have effective drying capacity in a rapid manner.
- Its solubility should be least in the solvents.
- It should not promote the reactions like polymerization, condensation, or auto-oxidation catalytically.
- It should work out in an economical way.

The most commonly used drying agents for organic solvents are:

Alcohols: Anhydrous potassium carbonate (K_2CO_3), calcium sulphate ($CaSO_4$), magnesium sulphate ($MgSO_4$), calcium oxide (CaO).

Alkyl & aryl halides: Anhydrous calcium chloride ($CaCl_2$) $CaSO_4$, $MgSO_4$, phosphoric oxide.

Aromatic hydrocarbons & ethers: Anhydrous $CaCl_2$, $CaSO_4$, phosphoric oxide.

Aldehydes & ketones: Anhydrous $CaSO_4$, $MgSO_4$, sodium sulphate (Na_2SO_4), K_2CO_3.

Organic bases: Solid potassium hydroxide (KOH) or sodium hydroxide (NaOH), CaO, barium oxide (BaO).

Organic acids: Anhydrous- $CaSO_4$, $MgSO_4$, Na_2SO_4.

General Procedure of Drying:

- Solvent should be shaken with small amount of drying agent (excess amount is avoided to minimize loss by adsorption).

- If good amount of water is present, the aqueous phase may separate out which can be removed by filtering with water-repellant paper which retains the aqueous layer.

- The resulting solvent is again treated with fresh drying agent, left overnight, filtered and distilled under inert atmosphere.

- The common drying agents, their applications and practical limitations are presented below.

Anhydrous Calcium Chloride ($CaCl_2$):

- It is not only an agent with high drying capacity but also highly economical.

- Its water absorption is very good but the process is slow due to blanketing of the particles with a thin layer of solution formed by extraction of water present.

- Thus it should be kept in contact for prolonged period during which the water combines forming a solid lower hydrate which also acts as a desiccating agent.

- It contains small amount of calcium hydroxide and basic chlorides as impurities and thus cannot be used to dry acids or acidic liquids.

- It combines with alcohols, phenols, amines, aminoacids, amides, ketones and some aldehydes and esters thus cannot be used for drying these classes of compounds.

- It is best suited agent for alkyl and aryl halides, aromatic hydrocarbons and ethers.

Magnesium Sulphate ($MgSO_4$):

- Monohydrate form is the most effective one and fully hydrated form is hepta hydrate ($MgSO_4.7H_2O$).

- It is an excellent, neutral desiccant, rapid in action, chemically inert and fairly efficient.

- It is used for drying most compounds including those in which calcium chloride is not applicable like aldehydes, ketones, amides, esters, nitriles.

Anhydrous Sodium Sulphate (Na₂SO₄):

- It is an inexpensive, neutral drying agent with very high water adsorption capacity (ten times its weight).

- Its drying action is slow and incomplete, thus used for preliminary removal of large quantities of water.

- It is inefficient in drying solvents like benzene and toluene.

- It is useless at temperatures above 32°C as deca hydrate begins to loose water of crystallization.

Anhydrous Calcium Sulphate (CaSO₄):

- It is prepared by heating calcium dihydrate or hemihydrates in an oven at 230°C-240°C for 3 hours which is also available commercially with the name Drerite.

- This reagent is chemically inert, insoluble in organic solvents, with efficient and extremely rapid drying action.

- The major limitation is its limited absorption capacity as it converts into its hemihydrates form. It absorbs only 6.6% of its weight. The porous form may absorb upto 10% if its weight.

- Thus the solvents should be preliminarily dried with either magnesium or sodium sulphate before drying with calcium sulphate.

Anhydrous Potassium Carbonate (K₂CO₃):

- It is a reagent with moderate drying capacity and used as preliminary drying agent.

- It is used for drying nitriles, ketones, esters and few alcohols and cannot be used for acids, phenols and other acidic substances.

- It is used as a substitute for sodium hydroxide or potassium hydroxide whenever there is a need for avoidance of strong alkaline reagent.

- It also finds its application in salting-out of water soluble alcohols, amines and ketones.

- Magnesium sulphate is the substitute as desiccant for this reagent.

Sodium and Potassium Hydroxides (NaOH/KOH):

- These are used for drying of amines and potassium hydroxide is superior to sodium hydroxide.
- These reagents reacts with many organic compounds like acids, phenols, esters, amides and also with solvents like chloroform thus has limited applications.

Calcium Oxide (CaO):

- It is commonly used for drying of alcohols of low molecular weight whose efficiency can be increased when preheated to 700°C-900°C.
- Calcium oxides as well as calcium hydroxide are insoluble in solvents, thermostable and non-volatile.
- Its high alkalinity hinders its utility to dry acidic compounds and esters.

Phosphoric Oxide:

- It is an extremely efficient reagent with rapid action.
- However it is expensive, difficult to handle, smells badly and has tendency for forming a protective syrupy coating on its surface.
- Preliminary treatment with anhydrous magnesium sulphate is necessary before using phosphoric oxide.
- It is used only when extreme desiccation is required and is used for hydrocarbons, ethers, alkyl and aryl halides and nitriles.
- It is not suitable for alcohols, acids, amines and ketones.

Zeolites (Molecular sieves):

- The final traces of water are removed by using molecular sieves and stored under them. These are a group of dehydrated synthetic sodium and calcium aluminosilicate adsorbents called as zeolites.
- These are having a crystal lattice structure which incorporates uniformly sized pores and are able to accept the molecules smaller than limiting dimensions.
- Gases and liquids of water diffuse into the pores and get retained inside by strong adsorption.

- Four types of molecular sieves are available viz 3A, 4A, 5A and 13X with pore sizes of 0.3, 0.4, 0.5 and 1.0 nm respectively.
- These are stable in the pH ranges of 5-11 and thus interaction with strong acids should be avoided.
- AW 300 and AW400 molecular sieves are used for drying acids.
- The beauty of molecular sieves lies in their regeneration by heating in an oven at 150°C-350°C and cooling desiccators.

Purification and drying procedures of solvents:

Saturated aliphatic hydrocarbons:

Light petroleum, cyclohexane, methylcyclohexane, decalin.

Light Petroleum (Petroleum ether, Ligroin): The commonly used refined petroleum have boiling points 40°C-60°C, 60°C-80°C, 80°C-100°C and 100°C-120°C.

Chief impurities: Unsaturated and aromatic hydrocarbons as impurities.

Purification and drying procedure:

Method 1:

- Shake the solvent with 10% of the volume of concentrated sulphuric acid for 2-3 times.
- Continue shaking with concentrated solution of potassium permanganate in 10% concentrated sulphuric acid until the color of permanganate remains unchanged.
- Wash the resulting solvent thoroughly with sodium carbonate solution followed by water.
- Dry over anhydrous calcium chloride and distil. For perfect drying allow it to stand over sodium wire or calcium hydride.

Method 2:

- Treat the solvent with concentrated sulphuric acid.
- Decant the treated solvent directly on to Grade 1 Basic Alumina (50 g/100 ml solvent) column, discard first 5% of the eluate and collect the remaining.

- This method also ensures the removal of peroxides too.
- Spectroscopic grade solvents can be obtained by passing through the column having Grade I silica gel in lower section and grade 1 basic alumina in the upper section.

Cyclohexane, methylcyclohexane and decalin can also be purified by following the above method.

Aromatic Hydrocarbons:

Benzene, toluene, xylenes.

Benzene: (C_6H_6-Highly Carcinogenic)

Caution: All procedures using benzene should be carried out in a well ventilated fuming cupboard and protective gloves should be worn.

Chief Impurities: Thiophene

Purification and drying procedure

Thiophene (boiling point 84°C) cannot be separated by distillation or by fractional crystallization.

Test for thiophene: Shake benzene (3.0 ml) with a solution of 10 mg isatin in 10 ml concentrated sulphuric acid (H_2SO_4). Appearance of bluish green color indicates the presence of thiophene.

- Treat (stir mechanically for 20-30 minutes) with 15% of its volume of concentrated sulphuric acid and repeat the process with fresh lot of concentrated sulphuric acid until the acid layer is colorless or pale yellow or until thiophene test shows negative.
- Wash the resulting solvent with water followed by 10% sodium carbonate solution again with water and dry under calcium chloride.
- Filter, distill off and collect the fraction obtained in the range of 80°C-81°C.
- To obtain perfect dried condition, store the distilled benzene over either sodium wire or Type 5A molecular sieves.

Phosphorus pentoxide, lithium aluminum hydride or calcium hydride can be used in place of sodium wire.

Toluene contains methyl thiophene (boiling point 112°C-113°C) as impurity which can also be purified and dried in the similar manner to that of benzene.

Xylenes are purified and dried by following the similar procedures mentioned above.

Halogenated hydrocarbons: Dichloromethane, chloroform, carbon tetrachloride.

Dichloromethane (CH_2Cl_2, Methylene chloride)

Purification and drying procedure:

- Wash the solvent with portions of concentrated sulphuric acid until the acid layer remains colorless.

- Then wash with water followed by sodium carbonate solution (10%) and again with water.

- Initially dry over anhydrous calcium chloride and then distill from calcium hydride and store in a brown colored bottle and keep away from sunlight over Type 3A molecular sieves.

Chloroform ($CHCl_3$):

Caution: It is carcinogenic. Thus where ever possible it should be replaced by dichloromethane as an extraction solvent.

Chief impurities: 1% ethanol added as stabilizer is the major impurity.

Purification and drying procedure:

Method 1:
- Shake chloroform with half of its volume of water for 5-6 times.
- Dry over anhydrous calcium chloride for atleast 24 hours and distill.

Method 2:
- Shake chloroform with a small volume of concentrated sulphuric acid for 2-3 times.
- Wash with water thoroughly and dry it over either anhydrous calcium chloride or anhydrous potassium carbonate and distill.

Method 3:

- Pass chloroform through Grade1 basic alumina (10 g /14 ml of solvent) which also removes traces of water and acid in it.

Note:

Phosphorous pentoxide, calcium sulphate or powdered Type 4A molecular sieves can be used in place of anhydrous calcium chloride.

It must not be dried by standing with sodium as it results in explosion.

The alcohol free chloroform should be kept in dark to prevent photochemical formation of dangerous quantities of phosgene ($COCl_2$) gas.

Carbon Tetrachloride (CCl_4):

Caution: It is carcinogenic and thus inhalation of vapors and contact with skin/eyes to be avoided.

Chief Impurities: Carbondisulphide is the chief impurity. Its permitted content in analytical grade solvents is not more than 0.005 % and in technical grade solvent is not more than 4%.

Purification and drying procedure:

- Shake commercial grade carbontetrachloride with potassium hydroxide (1.5 times the quantity required to combine with carbondisulphide) dissolved in an equal weight of water and 100 ml rectified spirit vigorously for 30 minutes at 50°C-60°C.
- Wash with water and repeat the process with half the quantity of potassium hydroxide.
- Remove the ethanol by shaking with 500 ml of water.
- Shake the resulting solvent with small portions of concentrated sulphuric acid until there is no coloration.
- Then wash with water and dry it over anhydrous calcium chloride and distill.
- Pass the distilled solvent through column of alumina and allow standing over Type 5A molecular sieves for perfect drying.

It must not be dried by standing with sodium as it results in explosion.

Aliphatic Alcohols:

Methanol, Ethanol, 1-Propanol, 2-Propanol, higher alcohols.

Methanol (CH₃OH, Methyl alcohol):

Chief impurities: Water and acetone content is NMT 0.1% and 0.02% respectively.

Purification and drying procedure:
- Water can be removed by fractional distillation easily as it do not form constant boiling point mixture.
- By making the fractionally distilled solvent by standing over Type 3A molecular sieves or by treatment with magnesium metal results in anhydrous methanol.
- Removal of acetone (> 1.0%): Heating to reflux the mixture of methanol (500 ml), furfural (25 ml), sodium hydroxide (10%, 60 ml) for 12 hours forms a resinous matter which carries all the acetone. Then fractionate the resulting methanol through an efficient column, discard the first 5 ml which may contain formaldehyde and collect the remaining.

Ethanol (C₂H₅OH, Ethyl alcohol):
- High degree of purity (99.5% - Absolute alcohol) required for most of the preparative purpose.
- Usually it is mixed with methanol to make it unfit for consumption and is called as Industrial Methylated Spirit (IMS) or Industrial spirit (IS) which can be used for recrystalization purpose.

Purification and drying procedure:

The absolute alcohol can be obtained from rectified spirit (95%) as follows
- Cool the freshly ignited calcium oxide (500 g) and rectified spirit (2 litres) taken in a round bottomed flask in a desiccators and heat it to reflux for 6 hours under calcium guard tube.
- Allow it to stand overnight and distill of the ethanol gently under calcium guard tube, discarding the first 20 ml of the distillate.
- Store the collected alcohol in a closed container.

Super Dry ethanol: (99.8% pure)
- Place clean/dry magnesium turnings (5 g), iodine (0.5 g) and 75 ml absolute ethanol in a 2 L round bottomed flask fitted with a double surface condenser and calcium guard tube and warm until the iodine disappears.

- If no live evolution of hydrogen is seen, add 0.5 g of iodine again and continue heating until all the magnesium is converted into ethanolate then add absolute ethanol (900 ml) and reflux for 30 minutes.

- Distill off the ethanol directly into the vessel in which it should be stored under complete anhydrous condition and store it on Type 4A molecular sieve.

Propan-1-ol ($CH_3CH_2CH_2OH$):

Purification and drying procedure:

- Dried over anhydrous potassium carbonate or anhydrous calcium sulphate, distilled off and the fraction, boiling point $96.5°C$-$97.5°C/760$ mmHg is collected.

- Super dry propanol can be obtained by treating with magnesium as described in ethanol.

Propan-2-ol : ($CH_3CH(OH)CH_3$, Isopropyl Alcohol)

Available in market with 91% and 99% purity.

Chief impurities: Peroxides.

Test for Peroxides:

- Acidify the mixture of 0.5 ml of propan-2-ol and 1.0 ml of 10% potassium iodide with dilute hydrochloric acid (HCl) (1:5).

- Mix with few drops of starch just before the test. Appearance of blue or bluish black coloration within 1 minute indicates peroxides.

Purification and drying procedure:

- Reflux 1.0 L propan-2-ol and 15 g of solid tin chloride for 1 hour and test for peroxides.

- If blue coloration appears, add 5 g of tin chloride and heat it to reflux for another half an hour periods until the test is negative.

- To the resulting propan-2-ol, add 200 g of calcium oxide and heat it to reflux for 4 hours.

- Distil off the propan-2-ol, discarding the first portion and allow it to stand over calcium metal or Type 5A molecular sieves for several days.

Higher alcohols:

Purification and drying procedure:

Secondary-butyl alcohol (butan-2-ol) and tertiary-butyl alcohol (2-Methyl-propan-2-ol) are dried by allowing to stand on Type 3A molecular sieves for 6 hours.

Other higher alcohols can be dried by using either anhydrous potassium carbonate or anhydrous calcium sulphate and fractionated after filtration.

Absolute butan-2-ol can be obtained by the following procedure.

- Place butan-2-ol containing not more than 0.5% water (1.0 L) and sodium metal (7.0 g) in a round bottomed flask fitted with double surface reflux condenser and calcium guard tube and warm until all the sodium has reacted.
- Then add pure 2-butyl succinate (33.0 g) or 2-butyl phthalate (41.0 g) and heat for gentle refluxing for 2 hours.
- Distill through Vigreux column to give alcohol containing water not more than 0.05%.

Monoalkyl ethers of ethylene glycol and diethylene glycols:

- Monomethyl, ethyl and butyl ethers of ethylene glycol are called as methyl cellosolve, cellosolve and butyl cellosolve respectively.
- Monomethyl, ethyl and butyl ethers of diethylene glycol are called as methyl carbitol, carbitol and butyl carbitol respectively.

Purification and drying procedure:

- The chief impurity present in these solvents is peroxides which can be removed by following the procedure described for propan-2-ol or by passing through grade 1 activated basic alumina column.
- They can be dried over anhydrous potassium carbonate or anhydrous calcium sulphate, filter and fractionate.
- The boiling points of methyl cellosolve, cellosolve and butyl cellosolve are 124.5°C, 135°C and 171°C respectively at 760 mmHg.

Carbitols can also be dried and purified in the similar way as that of cellosolves.

Ethers:

Diethyl ether, di-isopropyl ether and dibutyl ether.

Diethyl Ether (Solvent Ether): Highly flammable and extremely volatile. It should never be heated in bare flame. Either a steam bath or an electrically heated system should be used.

Chief Impurities: Water and ethanol

When allowed to stand in contact with air exposing to sunlight, oxidation occurs to give a highly explosive diethyl peroxide.

Test for peroxides is similar to that mentioned in propan-2-ol.

Purification and drying procedure:

For removal of peroxides:

Method 1:

Shake ether (1.0 L) with 20 ml of concentrated solution of ferrous (II) salt [prepared by either dissolving ferrous sulphate in a mixture of concentrated sulphuric acid (6.0 ml) and water (110 ml) or by dissolving ferrous chloride (100 g) in a mixture of concentrated hydrochloric acid (42 ml) and water (85 ml)].

Method 2:

Shake ether either with solid tin chloride or aqueous solution of sodium sulphite or pass through a column of alumina.

For removal of water and ethanol:

Method 1

Divide ether in the bottle into 2 portions and shake in a large separating funnel with ferrous sulphate solution (20 ml of the above solution diluted with 100 ml of water), discard the aqueous solution, combine the ether extracts.

To the combined extracts add anhydrous calcium chloride (200 g) and allow it to stand for 24 hours with occasional shaking (most of water and ethanol removed during the process).

- Filter ether through a fluted filter paper into a clean bottle and press sodium wire (7.0 g) into ether using sodium press.

- Close the bottle with a rubber stopper carrying calcium guard tube to exclude moisture and permit the escape of hydrogen and left for 24 hours.

- If on the subsequent day no evolution of hydrogen is occurring and surface of the sodium wire is bright, bottle is to be closed with a stopper and stored in a dry place away from sunlight.

- If the surface of sodium is not bright the process should be repeated.

Method 2

- Heat ether (1.0 L) to reflux with sodium benzophenone ketyl (prepared by using 10 g of benzophenone, 10 g of sodium) under nitrogen atmosphere until the blue or purple color persists.

- Distill the pure ether immediately before use.

The peroxides present in di-isopropyl ether and dibutyl ether can also be removed in the similar way to that of ether.

They can be dried over anhydrous calcium chloride or calcium hydride or sodium metal.

Tetrahydrofuran (C_4H_8O):

The commercial grade tetrahydrofuran is 99.5% pure. Peroxides and water are the major impurities.

Purification and drying procedure:

- Peroxides are removed by passing through a column of alumina or by shaking with ferrous sulphate solution as described in light petroleum and ether respectively.

- If latter method is used the solvent should be dried initially over calcium sulphate or solid potassium hydroxide before heated to reflux under calcium hydride or lithium aluminum hydride.

- Solvent is finally fractionally distilled (65°C-66°C/760 mmHg) and stored by adding antioxidant like 2,6-di-t-butyl-4-methoxy phenol (0.025%) as stabilizer.

Dioxane: ($C_4H_8O_2$, 1,4-dioxan, Diethylene dioxide)

Dioxane is a very useful reagent for a variety of organic compounds. Its solvating properties are similar to or better than that of ether. It is miscible with water in all proportions and is extremely hygroscopic.

Chief impurities: The commercial grade dioaxane contains small quantities of acetaldehyde and appreciable amounts of ethylene glycol along with some amount of water.

Purification and drying procedure:

Commercial dioxane (1.0 L), concentrated hydrochloric acid (14.0 ml) and water (100. 0 ml) are heated to reflux in a fuming cupboard for 12 hours while passing a stream of nitrogen through dioxane to remove the formed acetaldehyde.

Cool the solvent and treat with excess of potassium hydroxide pellets with shaking until some remains undissolved and strongly alkaline aqueous layer is run off.

Keep the resulting dioxane over fresh potassium hydroxide for 24 hours to remove the residual water present.

Decant the solvent and reflux under sodium wire (excess) for 12 hours i.e. until reaction ceases and sodium remains bright.

Distill off and store in a bottle wrapped with black paper out of contact with air.

Ketones: Acetone, ethyl methyl ketone.

Acetone (CH_3COCH_3):

Chief impurities: methanol, acetic acid and water.

Purification and drying procedure:

Method 1:

- Heat acetone under reflux with successive quantities of potassium permanganate until violet color persists.
- Dry with anhydrous potassium carbonate or anhydrous calcium sulphate.
- Filter from the desiccant and fractionate.

Method 2:

- To acetone in a bottle (700 ml), add a solution of silver nitrate (3.0 g) in water (20 ml) followed by 1 M sodium hydroxide (20 ml) and shake for 10 minutes.
- Filter the mixture, dry with anhydrous calcium sulphate and distil.

Method 3:

Dissolve sodium iodide (finely powdered, 100 g) under reflux in boiling commercial acetone (440 g).

Cool the mixture in ice-water mixture to − 8°C and filter off the crystals formed and quickly transfer into a dry distilling flask.

Gently warm the solution and collect the distillate and store over a Type 4A molecular sieves.

Ethyl Methyl Ketone ($CH_3COCH_2CH_3$, Butan-2-one):

It is the solvent with similar properties as that of acetone except that it has higher boiling point, thus less inflammable.

Initial purification (for recrystalization purpose)

Treat the commercial product with either anhydrous potassium carbonate or anhydrous calcium sulphate, filter, fractionate through an efficient column and collect the fraction with boiling range of 79°C-80°C.

Final purification: (Bisulphite Method)

Shake the above solvent with excess of saturated sodium bisulphite solution until reaction ceases.

Cool to 0°C, filter off the bisulphate complex, drain fully, wash with little ether and dry in air.

Decompose the dry bisulphate complex with excess of sodium carbonate solution and distill in steam.

Salt out the ketone from the distillate with potassium carbonate, separate and dry with potassium carbonate (removes traces of carbon dioxide and sulphur dioxide) and filter.

Allow the solvent to stand for several hours over anhydrous calcium sulphate, distill off.

Sodium Iodide Method:

- Saturate with sodium iodide by boiling under reflux and filter off the solution through a hot water funnel.

- Cool in a freezing mixture, filter off the white crystals formed (melting point 73°C-74°C).

- Heat gently the crystals in a fractional distillation assembly and collect the pure ethyl methyl ketone.

Esters

Methyl acetate, ethyl acetate.

Methyl acetate (CH₃-O-CO-CH₃):

Anhydrous product of 99% purity is available commercially. It is appreciably soluble in water.

Purification and drying procedure:

- Heat methyl acetate (1.0 L) under reflux with acetic anhydride (85 ml) for 6 hours.
- Distil through a fractionating column and collect the solvent distilling in the range of 56°C-57°C.
- Shake the solvent with anhydrous potassium carbonate for 10 minutes, filter and redistill the 99.9% pure solvent.

Ethyl Acetate (CH₃-COO-C₂H₅):

The anhydrous solvent with a boiling point of 76°C-77°C is available commercially. The 95%-98% pure solvent contains water, ethanol and acetic acid.

Purification and drying procedure:

- Heat the mixture of ethyl acetate (1.0 L), acetic anhydride (100 ml) and concentrated sulphuric acid (10 drops) to reflux for 4 hours and fractionate.
- Collect the distillate and shake with anhydrous potassium carbonate (20-30 g), filter and redistill from calcium hydride to get the solvent with 99.7% purity.

Nitrogen Containing Solvents:

Formamide (HCONH₂):

It is an excellent solvent for many polar organic compounds and for section of inorganic salts.

It is highly hygroscopic and readily hydrolyzed by acids or bases.

Chief impurities: Formic acid, water and ammonium formate.

Purification and drying procedure:

- Pass ammonia gas into the solvent until a slight alkaline reaction is obtained and add dry acetone to precipitate ammonium formate.
- Filter the solution and dry over magnesium sulphate and fractionally distil under reduced pressure.
- Pure formamide has a boiling point of 105°C/11mmHg.

N, N-Diemthyl Formamide: (H-CO-N(CH₃)₂, DMF)

It is a widely used solvent for many recently developed synthetic procedures due to its powerful solvating properties and its chemical stability in the absence of acidic or basic catalysts.

It undergoes decomposition to various degrees during atmospheric distillation and when it comes in contact with desiccants like sodium or potassium hydroxide or calcium hydride.

Purification and Drying Procedure:

Method 1:

*   Distill a mixture of DMF (1.0 L) benzene (100 ml) at atmospheric pressure and collect the water:benzene azeotrope (constant boiling mixture) which distills between 70°C and 75°C.
*   Shake the residual solvent with powdered barium oxide or with grade 1 activated alumina, filter and distill under nitrogen at reduced pressure.
*   Collect the fraction having boiling point 76°C/39 mmHg or 40°C/10 mmHg and store over Type 4A molecular sieves for 72 hours.

Method 2:

*   Dry the DMF over anhydrous calcium sulphate or Type 3A molecular sieve for 72 hours.
*   Distill under reduced pressure.

N-Methyl pyrrolidone is also purified in the similar way as that of DMF.

Acetonitrile (CH₃CN):

This is the most versatile solvent used for many synthetic procedures.

Chief impurities: Water, acetamide, ammonium acetate and ammonia.

Purification and drying procedure:

*   Remove water by allowing it to stand over silica gel or Type 4A molecular sieves [potassium hydroxide causes decomposition; calcium sulphate and calcium chloride are inefficient].
*   To the resulting solvent, add calcium hydride (phosphorous pentoxide is more efficient) in portions until the evolution of hydrogen ceases.
*   Decant the solvent and fractionally distill at atmospheric pressure.

Pyridine:

It is a hygroscopic liquid and forms hydrate of boiling point 94.5°C. It causes impotency in men and hence the vapors of pyridine should not be inhaled.

All the reactions using pyridine should be conducted in an efficient fuming cupboard. Disposable gloves should be worn while handling this solvent.

Purification and drying procedure:

- Heat to reflux a mixture of pyridine (1.0 L) and solid sodium hydroxide (100 g) for 12 hours and distill through a fractionating column.
- Add redistilled pyridine (400 ml) to a reagent prepared by dissolving anhydrous zinc chloride (340 g) in a mixture of concentrated hydrochloric acid (210 ml) and absolute ethanol (1.0 L).
- A crystalline precipitate of the addition compound ($2C_5H_5NZnCl_2$ HCl) separates out with evolution of heat.
- Filter off under suction and wash with little absolute ethanol and recrystallize in absolute ethanol (680 g yield).
- Add concentrated sodium hydroxide solution (40%) and steam-distill the liberated base until the distillate is no longer alkaline to litmus.
- Treat the steam distillate with 250 g of solid sodium hydroxide, separate the upper layer and extract the aqueous layer with two 250 ml portions of ether.
- Combine the upper layer and ether extracts and dry with anhydrous potassium carbonate, remove the ether on water bath and distill the pyridine through a fractionating column.
- Again heat it to reflux over calcium hydride or sodium hydroxide or potassium hydroxide pellets or barium oxide and distill with careful exclusion of moisture.
- Store the solvent over calcium hydride or barium oxide or Type 4A molecular sieves.

Quinoline is very less used as solvent and is purified in the similar way as that of pyridine.

Nitro Benzene ($C_6H_5NO_2$):

It is the most versatile solvent and sometimes used for recrystallization of compounds which do not dissolve well in the common organic solvents.

Vapors are very toxic and recrystallization should be carried out under efficient fuming cupboard.

After collection of crystals, it should be washed with volatile solvents such as ethanol or ether to remove excess of nitrobenzene.

The major limitation of nitrobenzene is its pronounced oxidizing action at the boiling point.

Chief impurities: Dinitrobenzene, nitro toluenes, aniline.

Purification and drying procedure:

- To the technical nitrobenzene add dilute sulphuric acid and steam distil the nitro benzene.
- Separate the nitro benzene in the mixture and dry with calcium chloride and distill under reduced pressure.
- The solvent has a boiling point of $210°C/760$ mm Hg.

Sulphur Containing Solvents:

Carbon disulphide, dimethyl sulphoxide.

Carbon Disulphide (CS_2):

It is a blood and nerve poison. It is highly flammable.

Distillation should be carried out on water bath at $55°C$-$65°C$ and if it gets overheated it gets ignited.

Purification and drying procedure:

- Shake technical grade carbon disulphide with 3 portions of potassium permanganate solution (5.0 g /L) for 3 hours, twice for 6 hours with mercury and finally with mercury(II) sulphate solution (2.5 g/L).
- Then dry over anhydrous calcium chloride and fractionate on a water bath at $55°C$-$65°C$.

Dimethyl Sulphoxide:

It is a highly useful water miscible solvent for many synthetic procedures and for spectroscopic work.

It is hygroscopic and distillation under atmospheric pressure causes decomposition.

Purification and drying procedure:

- Allow DMSO to stand over freshly activated alumina or calcium hydride or barium oxide or calcium sulphate overnight.
- Filter the solvent, fractionally distill over calcium hydride under reduced pressure and store over Type 4A molecular sieves.

Part C

Syntheses of
Drug Intermediates and
Drug Candidates (API's)

Introduction

The medicinal chemist's earlier isolated active ingredients from various (Majority plant) sources and used them for treatment. Due to the increased demand in the market for drugs due to population explosion all over the globe and limited availability/supply of raw material for the isolation of drug constituents, resulted in imbalance. This situation gave an impetus to synthesize those medicinally active compounds in the laboratory utilizing the basic concepts of organic chemistry in a much easier and cost effective manner. This opened up a new era of synthetic medicinal chemistry.

Synthesis of medicinally active compounds requires lot of imagination and intuition along with a profound knowledge of reaction mechanisms and practical skills in performing the reactions. That's why the synthesis is considered both an art and science.

The most accomplished approach for the synthesis of any drug/drug intermediate is disconnection or retro synthetic or synthon approach. In this approach the given molecule is dissected into smaller fragments intuitively such that each fragment represents one of the starting materials/synthons. These synthons when reacted under suitable conditions results in the target compound. Thus with the intention of making the students understand this concept, in all the experiments, retrosynthetic analysis has been included.

Before performing the reaction, student should understand the type of reaction that the reactants (starting material/synthons) undergo and should have awareness of the mechanism of the reaction. Then he/she will be in a position to appreciate the applicability of the basic organic chemistry principles which he/she has studied in the theory. To brush up the basics of organic chemistry the reaction along with its mechanism has been represented and explained in the principle section of the experiments.

The procedures for carrying out the reaction and work up of the reaction have been explained in a lucid and unique manner which assists the student not only to follow but also understand and reproduce easily. The physical data and category of the drugs follow the procedure section. Last but not the least the references are given at the end of each experiment for the students interested to know more about the molecule.

All of us are aware that the chemicals we use for the synthesis are harsh to environment and causes pollution. In order to protect the ecosystem, environmentally benign chemicals are recommended wherever possible. This is Green Chemistry. Heating the reaction mixture with the aid of microwaves not only speeds up the reaction but also gives comparably products in higher yields. Thus it is felt worthwhile to include a few experiments utilizing microwave assistance and concept of green chemistry in this section.

In this section, all the drugs / drug intermediates synthesized in one step, two steps and three steps are categorized accordingly. Further reactions involving green chemistry and Microwave assisted synthesis has been represented separately.

Preparations involving one step synthesis

## 1

Preparation of Phenolphthalein

Aim: To prepare phenolphthalein from phenol and phthalic anhydride.

Apparatus: Conical flask, water bath, round bottomed flask, reflux condenser, thermometer.

Introduced the following scheme in Retrosynthetic Analysis:

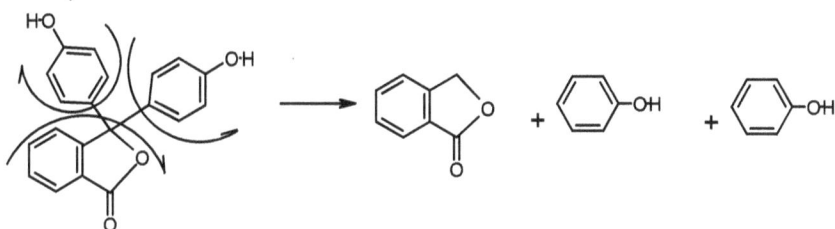

Phenolphthalein molecule can be split to remove 2 molecules of phenols which can act as one of the synthons for its preparation. To the remaining part of the molecule, if a carbonyl group is attached it results in phthalic anhydride. It can serve as the synthon as it undergoes electrophilic aromatic substitution with phenols to give the title compound.

Chemicals:

Chemical & Reagents	Mol. Wt.	Mol. Formula	Moles	Quantity Taken
Phthalic anhydride	148	$C_8H_4O_3$	0.012 M	1.7 g
Phenol	94	C_6H_6O	0.008 M	0.75 g (0.7 ml)
Concentrated sulphuric acid	98	H_2SO_4		0.5 ml
Ethanol	46	C_2H_5OH		*q.s.*

Principle:

Principle involved in the preparation of phenolphthalein is simple condensation reaction. Many phenols undergo condensation with phthalic anhydride to give useful products like cyclic anhydrides. Here phenolphthalein is prepared by condensing 2 moles of phenol with one mole of phthalic anhydride in the presence of a dehydrating agent like concentrated sulphuric acid (H_2SO_4) which on removal of a water molecule gives phenolphthalein.

As OH an electron releasing group it activates the ring system which enables the attack on positive carbonyl carbon centre of phthalic anhydride to give the intermediate which again is attacked by second phenolic moiety and undergo dehydration to give phenolphthalein.

Reaction:

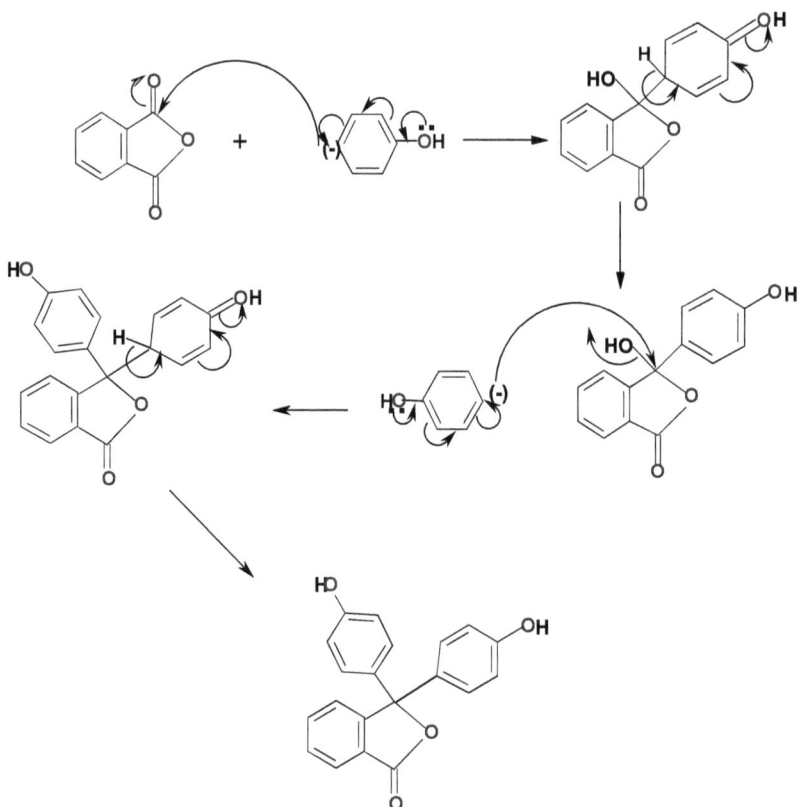

Procedure:

- Place phthalic anhydride (1.7 g, 0.012 M) and phenol (0.75 g, 0.7 ml, 0.008 M) in a 100 ml round bottomed flask and heat the reaction mixture on a sand bath until the solid melts.
- Then add 0.5 ml of concentrated H_2SO_4 and continue heating at 120°C for about 30 minutes under air condenser.
- Cool the reaction mixture to room temperature and add 30 ml of water mix thoroughly and filter the resultant precipitate at the pump.
- Wash the residue with water dry the product in an oven.
- Recrystallize the crude phenolphthalein from ethanol.

Physical Properties:

Melting Point: 261°C.

Percentage yield: 70% $^W/_W$

Category:

- It finds its use as laxative, acid base indicator in titrimetry, denaturant for industrial and laboratory alcohol.
- Used for the identification of ester group in organic compounds used in forensic sciences.

Identification tests:

- Does not give color with neutral ferric chloride ($FeCl_3$) solution.
- A drop of alcoholic solution turns sodium hydroxide (NaOH) solution to pink. But color disappears on standing or by addition of acid.

References:

1. Baeyer, A. *Ber.* **1871**, *4*, 658.
2. Baeyer, A. *Ann.* **1880**, *202*, 36.
3. Hubacher, M. US2192485 (1940 to Ex Lax).

2

Preparation of Phthalimide

Aim: To prepare phthalimide from phthalic anhydride and urea.

Apparatus: Conical flask, round bottomed flask, reflux condenser, thermometer.

Retrosynthetic Analysis:

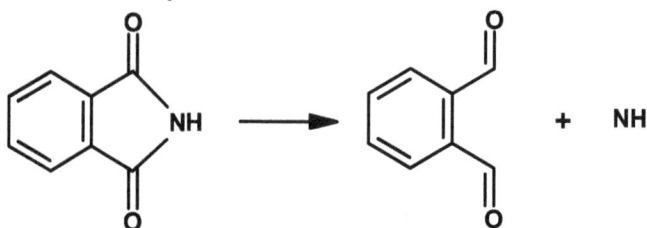

Splitting of phthalimide gives rise to amino and dibenzoyl groups. Hence ammonia can serve as amino group donor. However ammonia gas cannot be used directly thus urea can be utilized as the source of ammonia. To the remaining part of the molecule if the oxygen group is attached, it results in phthalic anhydride which can serve as another synthon. Thus by reacting phthalic anyhydride and urea we can get the target compound phthalimide.

Chemicals:

Chemicals & Reagents	Mol. Wt.	Mol. Formula	Moles	Quantity Taken
Phthalic anhydride	148	$C_8H_4O_3$	0.067 M	9.9 g
Urea	60	$C_1H_4N_2O$	0.03 M	2.0 g

Reaction:

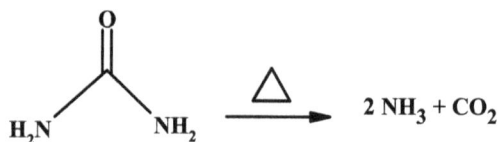

$$2 NH_3 + CO_2$$

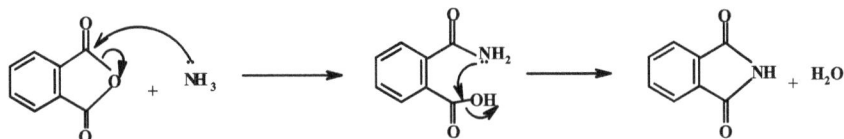

Principle:

The principle involved in the preparation of phthalimide is imide formation. Urea upon heating releases ammonia which then reacts with phthalic anhydride at high temperatures, leading to the formation of imide by loosing a molecule of water.

The lone pair of electrons on nitrogen of amino group attacks on the positive carbonyl carbon and gives corresponding 2-amido benzoic acid which again undergoes dehydrocyclization to give the phthalimide in quantitative yields.

Procedure:

- Place phthalic anhydride (9.9 g, 0.067 M) and urea (2.0 g, 0.03 M) in a 100 ml round bottomed flask and heat in an oil bath at 135°C under air condenser.

- The solid starts melting and effervescence commences gradually after 10-20 minutes. The mixture suddenly froths up above the original volume and is accompanied by rise in temperature to about 150°C-160°C.

- Discontinue the heating and cool the reaction mixture to room temperature.

- Add 10.0 ml of water to disintegrate the solid in the flask, filter the product formed at the pump, dry in an oven at 100°C and purify by recrystallization with industrial spirit.

Physical Properties:

Melting Point: 234°C.

Percentage yield: 85% $^W/_W$

Category:

- Phthalimide acts as an intermediate in the preparation of anthranilic acid.

Identification tests:

- **Phthalein test:**

  Phthalimide (0.1 g), phenol (0.1 g) and 2 drops of concentrated H_2SO_4 were fused together gently in a dry test tube. The mixture was cooled and water was added. To the resulting reaction mixture NaOH solution was added. Excess red coloration was produced which was decolorized by acid. It indicated the presence of phthalimide.

- **Fluorescein test:**

  Phthalimide (0.1 g), resorcinol (0.1 g) and 2 drops of concentrated H_2SO_4 were fused together gently in a dry test tube. Heating was continued until it is reddish brown in color. Reaction mixture was cooled and added water followed by NaOH solution in excess until alkaline. 1.0 ml of this solution was taken into another test tube and 10.0 ml of water was added. Green color fluorescence was observed.

References:

1. Noyes, W. A.; Porter, P. K. *Org. Syn.* **Coll. Vol. I,** 2nd Ed., pp 457.

3

Preparation of Para Amino Benzoic Acid

Aim: To prepare para amino benzoic acid from para nitro benzoic acid.

Apparatus: Beaker, condenser, round bottomed flask, funnel.

Retrosynthetic Analysis:

The p-amino benzoic acid can be split to give benzoic acid and ammonia. Both of these can serve as synthons. However ammonia cannot react with benzoic acid as both are electron rich species. Hence the alternative method is to take p-nitro benzoic acid as one of the synthons and reduce it by using reducing agents like Zn/HCl or Sn/HCl to give the target compound.

Chemicals:

Chemicals & Reagents	Mol. Wt.	Mol. Formula	Moles	Quantity Taken
p-Nitro benzoic acid	167	$C_7H_5O_4N$	0.010 M	1.8 g
Zinc powder	65	Zn	0.06 M	4.0 g
Concentrated Hydrochloric acid	36.5	HCl		9.4 ml
Ammonia solution	17	NH_3		q.s.
Glacial acetic acid	60	CH_3COOH		q.s.

Reaction:

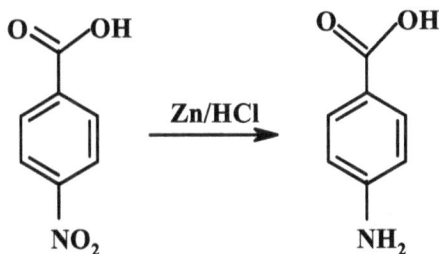

119

Principle:

The principle involved in the preparation of p-amino benzoic acid is reduction process. Aromatic nitro compounds can be reduced to amino compounds with metal and hydrochloric acid (HCl). Metals like zinc, copper and platinum are used for catalytic reduction.

Procedure:

- Place p-nitro benzoic acid (1.8 g, 0.010 M), zinc powder (4.0 g, 0.060 M) and concentrated hydrochloric acid (9.4 ml) in a 100 ml round bottomed flask fitted with a reflux condenser and heat the reaction mixture to reflux on a water bath.

- Shake the contents of the flask at regular intervals. (care should be taken that insoluble acid adhering to the flask is transferred into reaction mixture). After 20 minutes most of the zinc reacts and a clear solution is obtained.

- Cool the solution to room temperature and decant the supernatant liquid into a beaker. Wash the residual zinc with 10-15 ml of water and add washings into the beaker.

- Add ammonia solution until the mixture is just alkaline to litmus and digest the suspension on a steam bath for about 20 minutes.

- Stir the solution, filter at the pump and transfer into a beaker. Heat with 25 ml of water on a steam bath to ensure the extraction of the product and filter.

- Combine the filtrates, concentrate to half of the volume and filter if necessary to remove any precipitated salt.

- Acidify the filtrate with glacial acetic acid and evaporate on a water bath until the crystals commence to separate.

- Cool the solution, filter at the pump and dry.

Physical Properties:

Melting Point: 187°C.

Percentage yield: 75% $^W/_W$

Category:

- PABA is an intermediate in the synthesis of benzocaine.
- Antichromotrichia factor, anticanitic vitamin.

Identification tests:

- Soluble in hot water, alcohol and esters.
- Dissoved a pinch of sample in 1.0 ml of water and added 2-3 drops of saturated bleaching powder solution. A violet red solution developed.

References:

1. Nielson, J. *Chem. Soc.* **1962**, 371.

4

Preparation of Benzocaine

(Synonyms: Americaine, anaesthesin, flavamed)

Aim: To prepare benzocaine from p-amino benzoic acid(PABA).

Apparatus: Round bottom flask, reflux condenser, stand, water bath, funnel.

Retrosynthetic Analysis:

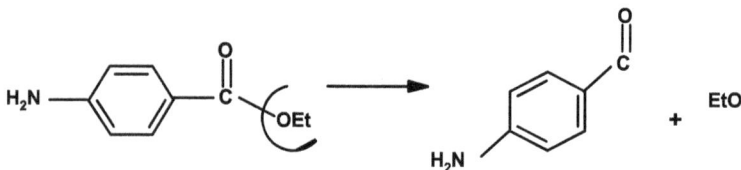

Splitting of benzocaine gives p-amino benzoyl and ethoxy groups. The p-amino benzoic acid and ethanol moieties can serve as p-amino benzoyl and ethoxy group donors. Thus these molecules can be used as synthons. This is an example for esterification of acids. It is carried out in the presence of acid as catalyst to make the reaction to move towards forward direction.

Chemicals:

Chemicals & Reagents	Mol. Wt.	Mol. Formula	Moles	Quantity Taken
Para amino benzoic acid	165	$C_7H_8O_2N$	0.006 M	1.0 g
Ethanolic HCl	82.5	$C_2H_5OH.HCl$		10.0 ml
Sodium carbonate	105	Na_2CO_3		q.s.
Sodium chloride	58	NaCl		q.s.

Reaction:

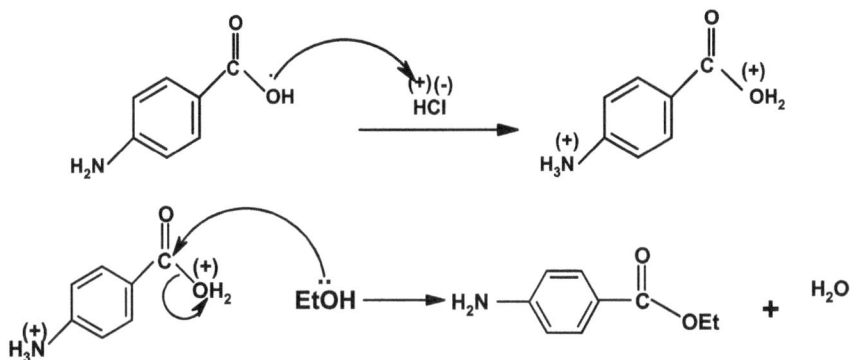

Principle:

The principle involved in the preparation of benzocaine is esterification reaction. Benzocaine is chemically ethyl para amino benzoic acid. It is used as a local anaesthetic. Para amino benzoic acid on reaction with ethanol in the presence of HCl gas forms the ester, benzocaine.

Procedure:

- Take p-amino benzoic acid (1.0 g) and saturated ethanolic HCl (10.0 ml) in a 100 ml round bottomed flask fitted with a condenser and calcium guard tube.
- Heat the reaction mixture to reflux on a water bath for 2.0 hours.
- Cool it to room temperature and pour into a beaker containing ice cold water.
- Neutralize the solution with solid sodium carbonate while cooling in an ice bath.
- Filter off the precipitated benzocaine, dry in a hot air oven and recrystallize from ethanol to give a pure product.

Preparation of saturated ethanolic HCl:

- Pass dry HCl gas through absolute ethanol until its weight increases by 25%-30% approximately.
- The dry HCL gas is produced by reacting concentrated H_2SO_4 with sodium chloride.

Physical Properties:

Melting Point: $88^\circ C$-$90^\circ C$;   Percentage yield: 54% $^W/_W$

Category:

- Used as an ointment to relieve pain associated with ulcers, wounds, burns and mucous surface and damaged skin (Local anaesthetic).
- Finds its use as lubricant and anaesthetic on intrathecal catheters, pharyngeal, nasal airways, nasogastric and endoscopic tubes.
- Used in otic preparations for temporary relief of ear pain.

Identification tests:

On a watch glass, 2-3 mg of compound was mixed with 4 - 5 drops of concentrated HCl. The solution was touched with a glass rod moistened with concentrated HNO_3. The solution turned green. When diluted with water and 1.0 ml of NaOH was added the solution turned orange.

References:

1. Salkowski, *Ber.***1895**, *28*, 1921.
2. Vorlander, Meyer. *Ann.* **1902**, *320*, 135.

## 5

Preparation of Acetoxime

Aim: To prepare acetoxime from acetone and hydroxylamine HCl.

Apparatus: Conical flask, beaker, spatula, water bath.

Retrosynthetic Analysis:

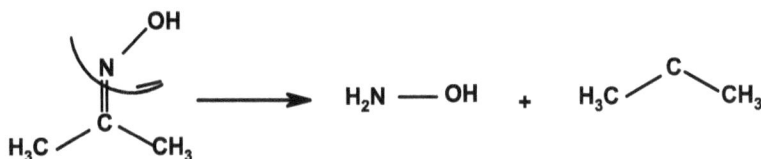

The splitting of acetoxime gives hydroxylamine as one of the synthons. Attachment of oxygen moiety to the remaining portion of the molecule gives another synthon, acetone. The combining of the 2 starting materials gives the title compound.

Chemicals:

Chemicals & Reagents	Mol. Wt.	Mol. Formula	Moles	Quantity Taken
Hydroxyl amine HCl	69.5	NH_4OCl	0.015 M	1.05 g
Acetone	58	C_3H_6O	0.016 M	0.95 g (1.2 ml)
Sodium hydroxide	40	NaOH	0.015 M	0.58 g
Cyclohexanone	78	C_6H_{12}		*q.s.*

Principle:

The principle involved in the preparation of acetoxime is oxime formation. Aldehydes and ketones condense rapidly with hydroxylamine to give crystalline oximes. Since hydroxylamine is usually available only in the form of salts, aqueous solution of this salt is treated with sodium acetate or sodium hydroxide to liberate base before treatment with aldehyde and ketone. Addition of hydroxylamine to acetone forms acetoxime by elimination of molecule of water. Sodium hydroxide should be weighed carefully because excess may dissolve the oxime as a sodium salt.

Reaction:

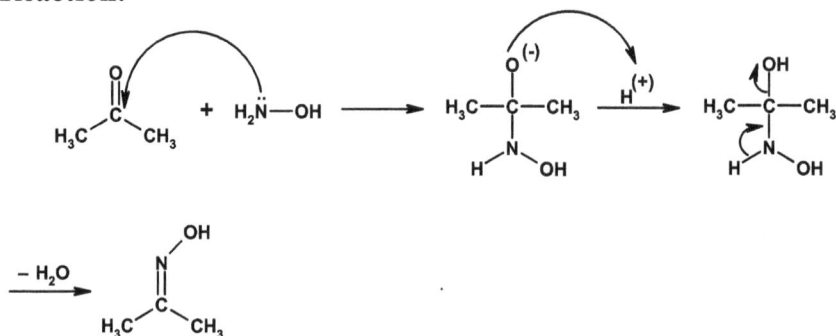

Procedure:

- Prepare a solution of hydroxylamine hydrochloride (1.05 g, 0.015 M) in water (2.0 ml) in a 25 ml beaker.
- Dissolve NaOH (0.58 g, 0.015 M) in water (1.0 ml) and cool in an ice bath.
- Add the ice cold solution of NaOH to the solution of hydroxylamine HCl by maintaining the temperature of the solution at 5°C-10°C.
- Add dry acetone (0.95 g, 1.2 ml, 0.016 M) drop wise to the above solution by gently shaking the reaction mixture in the ice bath.
- Acetoxime starts to crystallize when about half of the acetone is added.
- After complete addition of acetone, allow the mixture to stand for about 10-15 minutes at the same temperature.
- Filter off the crude product and without washing with other solvents, dry it in between folds of filter paper.
- Recrystallize from a minimum quantity of cyclohexanone to yield a pure product.

Physical Properties:

Melting Point: 59°C.

Percentage yield: 90% $^W/_W$

Category:

- To characterize the parent aldehyde or ketone.
- Some aldehydes and ketones are purified by converting into their oximes and hydrolyzing the later by boiling with acid.

Identification tests:

- Soluble in water, alcohol and ether.
- On boiling with HCl gives hydroxyl amine HCl.

References:

1. Semon, W. L. *Org. Syn.* **Coll. Vol. I**, 2nd Ed., **1941**, 318.

## 6

Preparation of Barbituric Acid

(Synonym: Malonyl urea.)

Aim: To prepare barbituric acid from urea and diethylmalonate.

Apparatus: Conical flask, round bottom flask, beaker, calcium chloride ($CaCl_2$), guard tube.

Retrosynthetic Analysis:

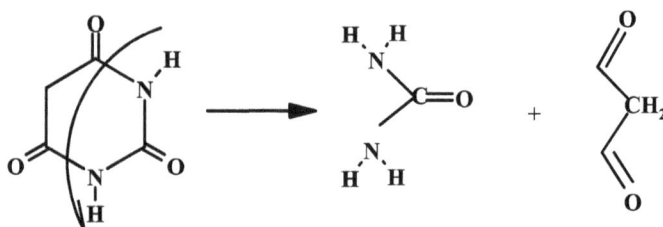

The splitting of barbituric acid gives rise to urea which is used as one of the synthons and a 3 carbon donor. A 3 carbon donor like diethylmalonate serves as another synthon. The combining of the above 2 molecules under strongly basic condition can give the title compound.

Chemicals:

Chemicals & Reagents	Mol. Wt.	Mol. Formula	Moles	Quantity Taken
Urea	60	$C_1H_4N_2O$	0.04 M	2.65 g
Diethylmalonate	160	$C_7H_{12}O_4$	0.03 M	5.0 g (4.7 ml)
Sodium metal	23	Na	0.03 M	0.65 g
Absolute ethanol	46	C_2H_5OH		15 ml
Anhydrous calcium chloride	111	$CaCl_2$		*q.s.*

Principle:

Principle involved in the preparation of barbituric acid is simple condensation of diethyl malonate with urea in the presence of a strong base, sodium ethoxide. The use of substituted derivatives of urea and diethyl malonate gives wide range of substituted barbiturates which finds use as central nervous system (CNS) depressants. Barbituric acid may be regarded as 2,4,6-trihydroxy derivative but in crystalline from it exists as triketo from.

Reaction:

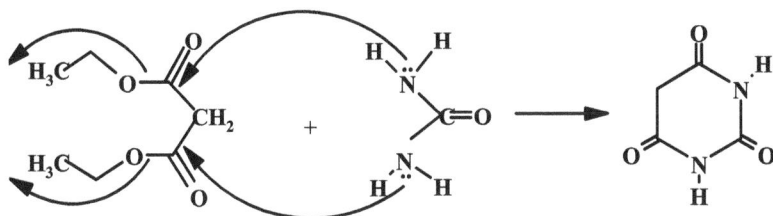

Procedure:

- Dissolve the sodium metal (0.65 g 0.03 M) in absolute ethanol (15.0 ml) taken in a 50 ml round bottomed flask fitted with a reflux condenser and calcium chloride guard tube.
- Add the diethylmalonate (5.0 g, 4.7 ml, 0.03 M) into the above reaction mixture.
- Meanwhile dissolve urea (2.65 g, 0.04 M) in minimum amount of boiling ethanol (5.0 ml) and immediately add to the flask containing diethylmalonate.
- Heat the reaction mixture to reflux at 110°C for seven hours. Sodium salt of barbituric acid rapidly precipitates out.
- Treat the reaction mixture with 45.0 ml of hot water and acidify with concentrated H_2SO_4.
- Keep aside the resulting clear solution overnight in the refrigerator.
- Filter off the precipitate obtained, dry in a hot air oven and recrystallize from hot water and ethanol.

Physical Properties:

Melting Point: 248°C.

Percentage yield: 70% $^W/_W$

Category:

- Derivatives of barbituric acids are CNS depressants used as sedatives and hypnotics.
- Used in manufacture of plastics and pharmaceuticals.

Identification tests:

- To few mg of compound in water, silver nitrate solution was added. Yellow color precipitate formed indicated the presence of barbituric acid.
- Barbituric acid alone when heated liberated ammonia.

References:

1. Dickey, G. *Org. Syn.* **Coll. Vol. II**, 1st Ed., **1943**, 60.

## 7

Preparation of Benzimidazole

Aim: To prepare benzimidazole from o-phenylenediamine.

Apparatus: Heating mantle, beaker, conical flask, funnel.

Retrosynthetic Analysis:

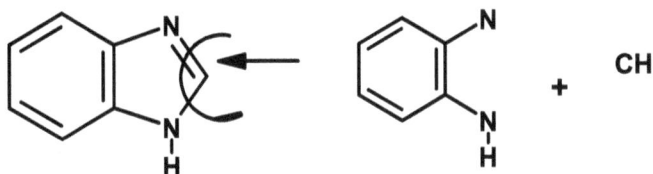

The benzimadazole moiety can split into o-phenylenediamine (OPDA) and methine group. The formic acid or formaldehyde can serve as a source of methine group. Thus OPDA and either formic acid or formaldehyde can be used as synthons for the preparation benzimidazole. However formic acid is preferred over formaldehyde as it can undergo dehydrocyclization easily with OPDA when heated under neat condition.

Chemicals:

Chemicals & Reagents	Mol. Wt.	Mol. Formula	Moles	Quantity Taken
o-Phenylenediamine	108	$C_6H_8N_2$	0.025 M	2.75 g
Formic acid (90 %)	46	CH_2O_2	0.04 M	1.75 g
Sodium hydroxide (10 %)	40	NaOH		*q.s.*

Principle:

Benzimidazole is obtained by the reaction between o-phenylenediamine and formic acid. It is an example for dehydrocyclization by heating the reaction mixture in neat condition without any solvent (fusion reaction).

Benzimidazole is a heterocycle which can act as a synthon for preparation of several drug substances like metronidazole.

Reaction:

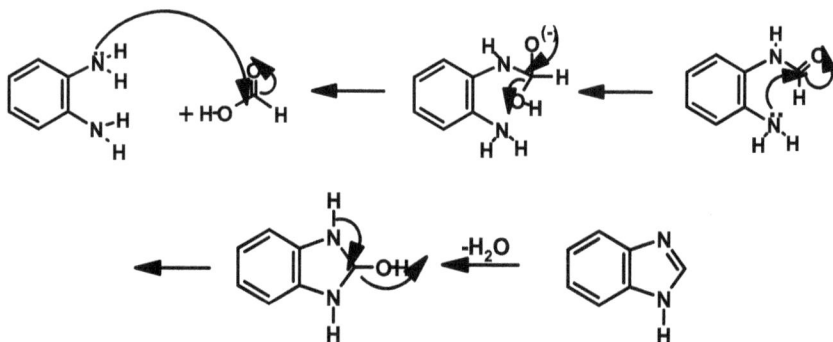

Procedure:

- Take an accurately weighed amount of o-phenylenediamine (2.70 g, 0.025 M) and formic acid-90 % (1.75 g, 0.04 M) in a round bottomed flask (50.0 ml).

- Heat the reaction mixture at 100°C for 2 hours.

- Cool the mixture and neutralize by sodium hydroxide solution (10%). Add little excess of sodium hydroxide to make it alkaline to litmus.

- Filter off the crude precipitate obtained, wash with cold water and dry.

- To the crude product, add boiling water (400.0 ml) and decolorizing carbon (2.0 g) and heat for 15 minutes to digest the precipitate.

- Filter the resulting solution through a preheated Buchner flask and cool the filtrate to 10°C.

- Separate the crystals formed by filtration and dry at 100°C.

Physical Properties:

Melting Point: 171°C-172°C.

Percentage yield: 85% $^W/_W$

Category:

- Serves as a synthon (starting material) for preparation of various drug substances.

References:

1. Wundt, *Ber.***1878**, *11*, 826.
2. Wagner, M. *Org. Syn.* **1943**, *Vol. II*, 65.

8

Preparation of Benzotriazole

Aim: To prepare benzotriazole from o-phenylenediamine.

Apparatus: Beaker, separating funnel, magnetic stirrer, ice bath, Buchner flask and funnel.

Retrosynthetic Analysis:

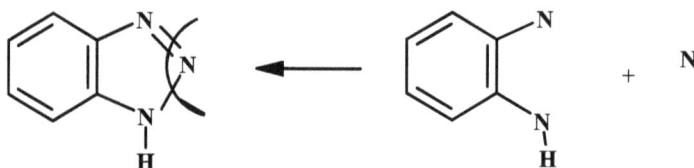

The benzotriazole moiety can split into o-phenylenediamine (OPDA) and nitrogen group. The nitrogen group should be introduced into OPDA. This can be achieved by taking OPDA as one of the synthons and diazotizing one of its amino group by sodium nitrite and acetic acid. This diazotized compound can undergo cyclization by the internal attack by free amino group onto diazonium nitrogen to give 2-hydroxy benzotriazole which in turn can loose water molecule to give the target compound.

Chemicals:

Chemicals & Reagents	Mol. Wt.	Mol. Formula	Moles	Quantity Taken
o-Phenylenediamine	108	$C_6H_8N_2$	0.1 M	10.8 g
Sodium nitrite	69	$NaNO_2$	0.11 M	7.5 g
Glacial acetic acid	60	$C_2H_4O_2$	0.2 M	12.0 g

Principle:

Benzotriazole is prepared by diazotization of amine group of o-phenylenediamine (OPDA). Sodium nitrite in the presence of glacial acetic acid at 0°C-5°C gives rise to nitrous acid which inturn reacts with amino group of OPDA to give diazonium salt which undergo cyclization by the attack of the lone pair of electrons from the neighboring amino group of OPDA with the loss of water molecule.

Reaction:

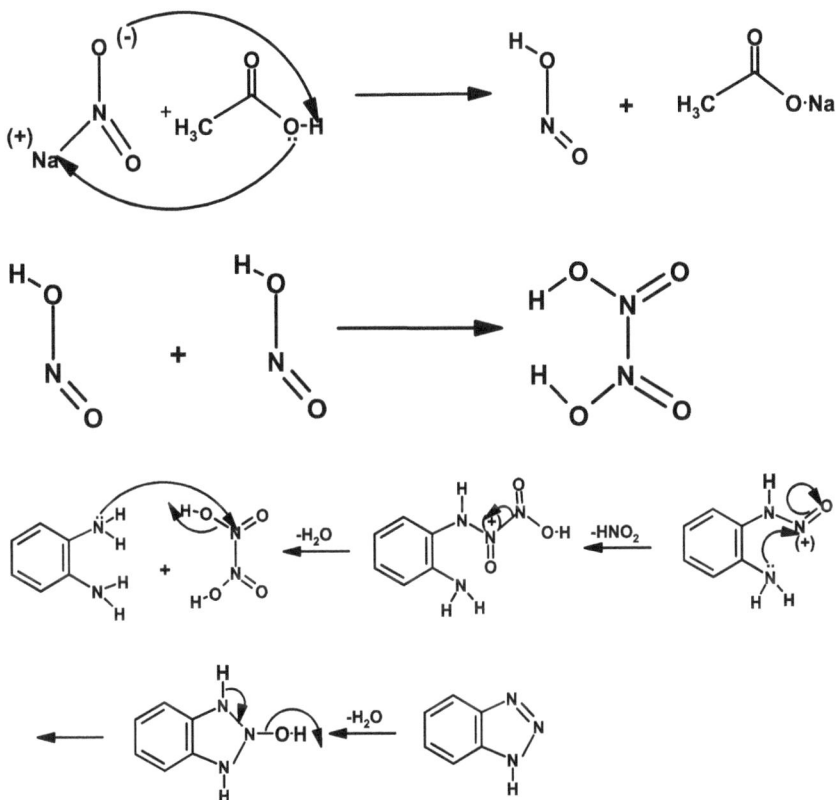

Procedure:

- Take o-phenelenediamine (10.8 g, 0.1 M), glacial acetic acid (12.0 g, 0.2 M), water (30.0 ml) in a beaker (250.0 ml) and dissolve by warming slightly if necessary. Cool the above solution to 15°C and stir on a magnetic stirrer.

- To the stirred solution, add sodium nitrite (7.5 g, 0.11 M) in 15.0 ml of water in one portion.

- The reaction mixture warms up within 2-3 minutes and the temperature rise to around 85°C and begins to cool. During this process color changes from deep red to pale brown.

- Continue the stirring for 15 minutes by which time the temperature drops down to 35°C-40°C. Chill the reaction mixture thoroughly in an ice water bath for 30 minutes.

- Filter the pale brown precipitate formed by vacuum filtration, wash with 30.0 ml portions of ice cold water several times and dry.

- Dissolve the solid obtained in around 150.0 ml of boiling water and decolorize by activated charcoal.

- Filter while hot and allow the filtrate to cool to room temperature and further cool in an ice bath. The pale yellow crystals are formed.

- Filter the separated crystals, dry and purify by recrystallization using benzene.

Physical Properties:

- Melting Point: 99°C, Percentage Yield: 70% $^W/_W$

References:

1. Ladenburg, *Ber.* **1876**, *9*, 219.
2. Damschroder, Peterson. *Org. Syn.* **1955**, *Coll. Vol. III*, 106.

9

Synthesis of 2,3-Diphenyl Quinoxaline

Aim: To prepare 2, 3-diphenyl quinoxaline from o-phenylenediamine.

Apparatus: Beaker, funnel, conical flask.

Retrosynthetic Analysis:

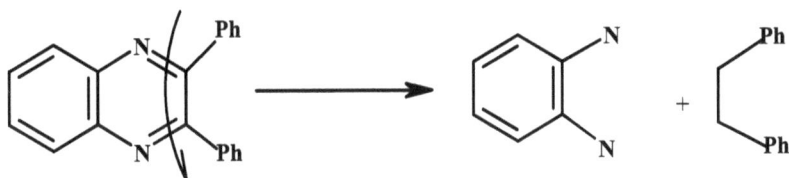

The splitting of 2,3-diphenyl quinoxaline gives ortho phenylene diamine (OPDA) and 1,2-diphenyl ethane. OPDA can be used as one of the synthons directly but 1,2-diphenyl ethane cannot react with OPDA due to lack of functional groups. Introduction of carbonyl functions into 1,2-diphenyl ethane gives benzil which can react with OPDA. Thus benzil can be used as another synthon. Both the synthons can react to undergo dehydrocyclization to give the target compound.

Chemicals:

Chemicals & Reagents	Mol. Wt.	Mol. Formula	Moles	Quantity Taken
o-Phenylenediamine	108	$C_6H_8N_2$	0.01 M	1.08 g
Benzil	210	$C_{14}H_{10}O_2$	0.01 M	2.10 g
Rectified Spirit	46	C_2H_5OH		15 ml

137

Principle:

The reagents like o-phenylenediamine and 1,2-dicarbonyl compound, benzil undergo cyclization, when heated to give quinoxaline compound.

Reaction:

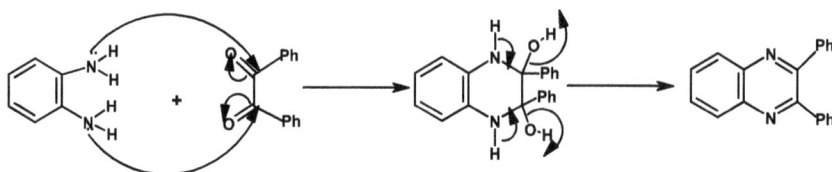

Procedure:

- Dissolve benzil (1.08 g, 0.01 M) and o-phenylenediamine (2.10 g, 0.01 M) in 8.0 ml of rectified spirit respectively in 2 separate beakers.

- Mix both the solutions and warm in a water bath for 30 minutes.

- To the resulting reaction mixture, add water until a slight cloudiness appears and allow it to cool.

- Filter the precipitate formed, dry and purify by recrystallization from aqueous ethanol.

Physical Properties:

Melting Point: 126°C.

Percentage yield: 80% $^W/_W$

References:

Vogel, A. I. Text book of Practical Organic Chemistry. Addison Wesley Longmann Ltd. England. 5th Ed. **1989**, PP 1184-1190.

Arun, S. *Systematic lab experiments in Organic Chemistry*. New Age International (P) Ltd. Publishers, New Delhi, 2nd Ed., **2006**, PP 110.

## 10

Preparation of Phenothiazine

Aim: To prepare phenothiazine from diphenyl amine.

Apparatus: Beaker, condenser, round bottomed flask, funnel, measuring jar.

Retrosynthetic Analysis:

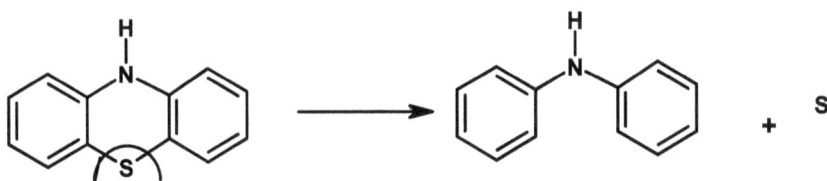

Phenothiazine can split into diphenylamine and sulphur. Thus diphenylamine and elemental sulphur can be taken as synthons and made to react in the presence of mild oxidizing agent like iodine to give phenothiazine.

Chemicals:

Chemicals & Reagents	Mol. Wt.	Mol. Formula	Moles	Quantity Taken
Diphenyl amine	169	$C_{12}H_{11}N$	0.01 M	1.69 g
Sulphur	32	S	0.02 M	0.64 g
Iodine crystals	254	I_2		Traces
Benzene	78	C_6H_6		q.s.
Petroleum ether				q.s.

Reaction:

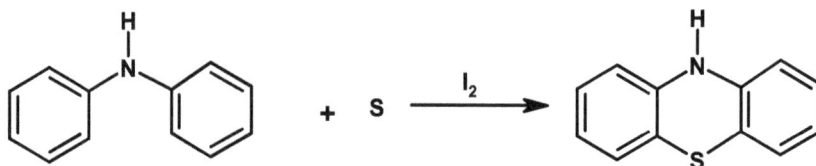

Principle:

Phenothiazine is an antipsychotic agent and its derivatives such as chlorpromazine are used in the treatment of schizophrenia. Phenothiazine is synthesized from diphenylamine and a mixture of sulphur and trace amounts of iodine.

Procedure:

- Take diphenylamine (1.69 g, 0.01 M),precipitated sulphur powder (0.64 gm, 0.02 M) and traces of iodine crystals in a round bottomed flask fitted with a reflux condenser and heat under reflux on an oil bath at a temperature of 195°C for 15 minutes.
- Adjust the temperature to 185°C and then heat the mixture for 3 hours.
- After 3 hours green solid product is obtained. The obtained product is insoluble in petroleum ether.
- The insoluble precipitated product is washed with petroleum ether and alcohol.
- Then extract the product using benzene and distill off the benzene to get dark colored phenothiazine as crystalline solid.

Physical Properties:

Melting Point: 185°C.

Percentage yield: 92% $^W/_W$

Category:

- Insecticide and antihelmentic.

References:

1. Knoevenegel, *J. Prakt. Chem.* **1914**, *89*, 11.

11

Preparation of Niclosamide

Aim: To prepare niclosamide from 5-chloro salicylic acid and 2-chloro 4-nitro aniline.

Apparatus: Conical flask, water bath, round bottomed flask, reflux condenser, thermometer, buchner funnel, electric water bath.

Retrosynthetic Analysis:

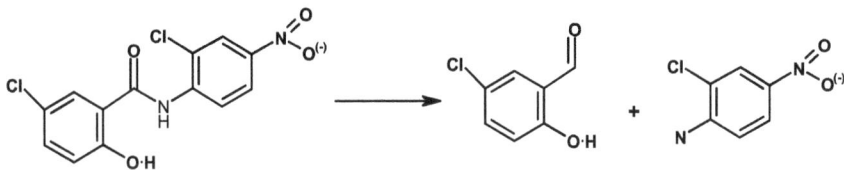

Niclosamide can be fragmented into 1-carbonyl-2-hydroxy-5-chloro benzene and 2-chloro 4-nitro aniline. The addition of hydroxyl group to carbonyl derivative results in 5-chloro salicylic acid. Addition of two hydrogens on to nitrogen of aniline results in 2-chloro-4-nitro aniline. Thus these two molecules serve as synthons.

Chemicals:

Chemical & Reagents	Mol. Wt.	Mol. Formula	Moles	Quantity Taken
5-chloro salicylic acid	156.5	$C_7H_5O_2Cl$	0.011	1.725 g
2-chloro-4-nitroaniline	158.5	$C_6H_5NO_2Cl$	0.013	2.09 g
Xylene	106	C_8H_{10}		25 ml
Phosphorous trichloride	137.5	PCl_3	0.0036	0.5 g
Ethanol	46	C_2H_6O	*q.s*	*q.s.*

Principle:

It is a simple dehydration reaction between carboxylic acid and an amine using phosphorus trichloride (PCl_3) as dehydrating agent and xylene as solvent. Nicosamide finds its use as an antihelmenthic.

Reaction:

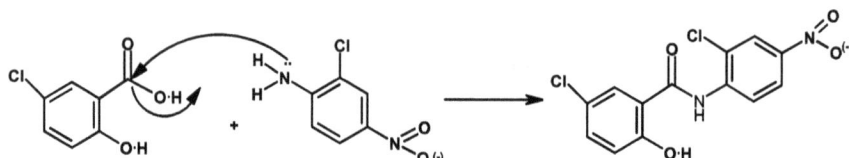

Procedure:

- Dissolve 5-chlorosalicylic acid (1.725 g, 0.011 M) and 2-chloro-4-nitroaniline (2.09 g, 0.013 mole) carefully in 25.0 ml of pure xylene in a round bottom flask fitted with a double surface condenser.

- Boil the reaction mixture on a heating mantle and introduce pure PCl_3 (5.0 g, 0.0036 M) in small lots, at regular intervals from the top end of the condenser. Continue the heating for further 2 hours.

- Cool the reaction mixture. Then the crude crystals of niclosamide starts separating out.

- Filter the crude product in a buchner funnel under suction, dry and recrystallize the crude product from ethanol.

Physical Properties:

Melting Point: 233°C.

Percentage yield: 80% $^W/_W$

Category:

- It is a potent anthelminthic especially effective against the *cestodes* that infect humans.

References:

1. Raghu Prasad, M., Raghuram Rao, A. Synthesis of niclosamide. Unpublished results.

12

Preparation of Fenbufen from Biphenyl

Aim: To prepare fenbufen from biphenyl.

Apparatus: Round bottomed flask, magnetic stirrer, calcium guard tube, steam distillation apparatus, Buchner funnel.

Retrosynthetic Analysis:

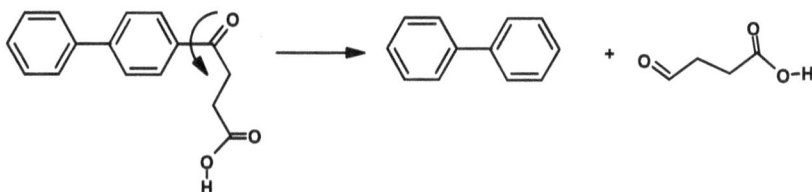

Fenbufen can be split into biphenyl and 1-carboxy ethane carbonyl fragments. Addition of OH group to carbonyl function of 1-carboxy ethane carbonyl gives rise to ethane-1,2-dicarboxylic acid. As it has 2 carboxylic acids which are equally reactive it cannot serve the purpose. Dehydration of this molecule gives rise to succinic anhydride which can react with biphenyl under Friedel Craft's reaction condition. Thus succinic anhydride and biphenyl can serve as synthons.

Chemicals:

Chemical & Reagents	Mol. Wt.	Mol. For.	Moles	Quantity Taken
Biphenyl	154	$C_{12}H_{10}$	0.02 M	3.08 g
Succinic anhydride	100	$C_4H_4O_3$	0.02 M	2.00 g
Aluminum chloride (Anhydrous)	133.5	$AlCl_3$	0.02 M	2.70 g
Hydrochloric Acid (15 %)	36.5	HCl		20.0 ml
Sulphuric acid (5 %)	98	H_2SO_4		*q.s.*

143

Principle:

This is a reaction similar to that of Friedel Craft's reaction. Biphenyl reacts with succinic anhydride in the presence of aluminum chloride as lewis acid, to yield p-phenylbenzoyl propionic acid, fenbufen. As phenyl substituent on benzene ring system is electron donating and ring activating. It directs the electrophile towards para position. Succinic anhydride acts as an electrophile. When electrons of benzene ring attacks, on the electron deficient carbonyl carbon of succinic anhydride, ring opens up to give propionic acid carbonyl side chain. As biphenyl is electron rich species reaction takes place very easily at $0°C$-$10°C$. Nitro benzene serves as a solvent.

Reaction:

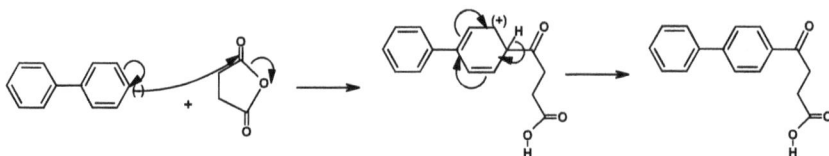

Procedure:

- Take aluminum chloride (2.70 g, 0.02 M) and dissolve in nitro benzene (10.0 ml) in a round bottomed flask fitted with a calcium guard tube and stir by cooling the mixture in an ice bath to $5°C$-$10°C$.

- Add a finely powdered mixture of biphenyl (3.08 g, 0.02 M) and succinic anhydride (2.00 g, 0.02 M) to the stirred solution of aluminum chloride while maintaining the temperature of reaction mixture below $10°C$.

- After complete addition, stir the reaction mixture at room temperature for 96 hours.

- Pour the resulting mixture into a solution of 20.0 ml of concentrated hydrochloric acid (prepared by 15 ml concentrated HCl in 100 ml of water) with constant stirring.

- Remove the excess of nitro benzene from the acidified mixture by steam distillation and filter the solid residue obtained under suction pump.

- Dissolve the solid in 3.0% hot sodium carbonate solution and reprecipitate by adding dilute sulphuric acid for purification.
- Carry out further purification by minimum amount of ethanol.

Physical Properties:

Melting point: 185°C-187°C.

Percentage yield: 75% $^W/_W$

Category:

- It is a nonsteroidal antiinflammatory drug prescribed in the treatment of pain and inflammation in rheumatic arthritis, osteoarthritis and ankylosing spondilytis.

Reference:

1. Hey, D. H.; Wilkinson, R. *J. Chem. Soc.* **1940**, 1030.
2. Weizmann, M. *et al., Chem & Ind.* **1940**, 402.
3. Reppe, W. Ann. **1955**, *596*, 223.

## 13

Preparation of Busulfan

Aim: To prepare busulfan from 1,4-butanediol.

Apparatus: Beaker, Buchner funnel, glass rod etc.

Retrosynthetic Analysis:

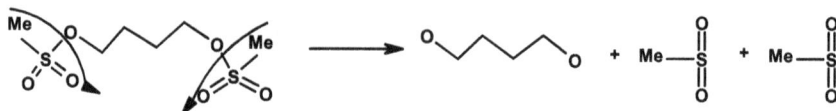

Busulfan can be split into 1,4-oxo butane and methane sulphur dioxide. Addition of hydrogens to oxygen function of fragment one gives rise to 1,4-butanediol. Chlorine addition to methane sulphur dioxide results in methane sulphonyl chloride. Thus both these molecules can serve as synthons.

Chemicals:

Chemical & Reagents	Mol. Wt.	Mol. Formula	Moles	Quantity Taken
1,4-Butane diol	90	$C_4H_{10}O_2$	0.01 M	0.9 g
Methane sulphonyl chloride	114	CH_3SO_2Cl	0.08 M	2.28 g
Pyridine	91	C_6H_5N		5.0 ml
Acetone	58			15.0 ml
Ether				15.0 ml

Principle:

This reaction is an example for O-sulphonation. The lone pair of electrons on the oxygen of the hydroxyl group present in butane diol attacks the positive sulphur and chlorine moves away as hydrogen chloride. 2 molecules of methane sulphonyl chloride are consumed by 2 hydroxyl groups. Pyridine is used as the solvent as it reacts with the HCl, the byproduct formed in the reaction to give pyridine hydrochloride salt which inturn helps the reaction to move forward direction.

Reaction:

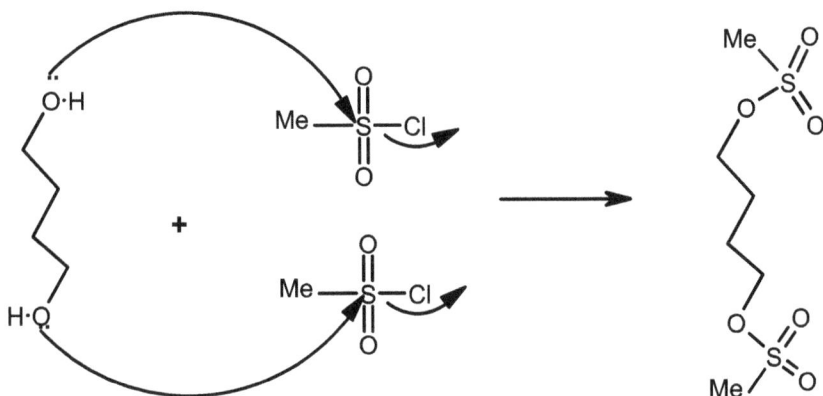

Procedure:

- Dissolve the redistilled 1,4-butanediol (0.9 g, 0.01 M) in 5.0 ml of redistilled pyridine taken in a round bottomed flask and stir it in an ice bath by cooling it to below 18°C.

- Add methane sulphonyl chloride (2.28 g, 0.02 M) drop wise to the above solution with constant stirring at such a rate that its temperature do not arise above 18 ± 2°C.

- After complete addition of methane sulphonyl chloride, continue the stirring of reaction at room temperature for half an hour more.

- As it is an exothermic reaction temperature will be raised to around 60°C by the end of 30 minutes and a thick precipitate of pyridine hydrochloride will be formed.

- Cool the mixture in an ice bath and add around 25.0 ml of ice cold water to dissolve the pyridine hydrochloride formed.

- Filter of the white crystalline product formed under vacuum pump, wash with ice cold water, dry and recrystallize using acetone ether mixture (1:1).

Physical Properties:

Melting Point: 115°C-116°C.

Percentage yield: 70% $^W/_W$

Category:

- It is an antineoplastic agent belonging to alkylating class of anticancer agents used in the treatment of chronic granulocytic leukemia, primary thrombocytosis and polycythemia.

References:

1. Haddow, A.; Timmis, G. M. *Lancet*, **1953**, *1*, 207.
2. Timmis, G. M. *US 2917432*.

## 14

Preparation of Chlorpropamide

Aim: To prepare chlorpropamide from p-chloro benzene sulphonamide.

Apparatus: Beakers, stirring rod, Buchner funnel etc.

Retrosynthetic Analysis:

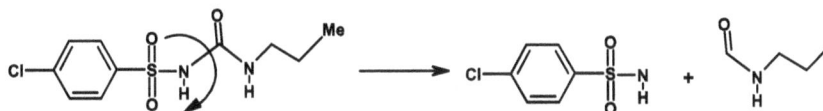

Chlorpropamide can be split into p-chloro benzene sulphonamido and propyl isocyanato fragments. Thus p-chlorobenzene sulphonamide and propyl isocyanate can act as synthons for the preparation of chlorpropamide.

Chemicals:

Chemical & Reagents	Mol. Wt.	Mol. For.	Moles	Quantity Taken
p-Chloro benzene sulphonamide	191.5	$C_6H_6NSO_2Cl$	0.01 M	1.91 g
Propyl isocyanate	75	C_4H_7NO	0.013 M	0.975 g
Dimethyl formamide (anhydrous)	73	C_3H_7NO		5.0 ml
Triethylamine (anhydrous)	101	$C_6H_{15}N$	0.05 M	5.05 g
Acetic acid (20.0 %)	60	CH_3COOH		200.0 ml

Principle:

The lone pair of electrons present on the nitrogen of sulphonamide attacks the positive carbonyl carbon present in the propyl isocyanate to give urea derivative, chlorpropamide. Here aprotic solvent, dimethyl formamide is used which helps to hasten up the reaction along with triethylamine which acts as a catalyst by providing basicity to the reaction medium. It is a simple condensation reaction in which active hydrogen atom from the sulphonamide function gets attached to nitrogen atom of the isocyanate without loss of any separate entity.

Reaction:

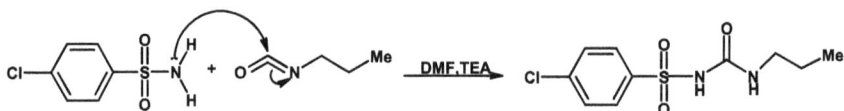

Procedure:

- Mix a pure solution of propylisocyanate (0.975 g, 0.013 M) dissolved in 5.0 ml of anhydrous dimethyl formamide.

- Suspend p-chloro benzene sulphonamide (1.91 g, 0.01 M) in triethylamine (5.05 g, 0.05 M).

- Cool it in an ice bath to below 10°C and stir vigorously.

- Add propylisocyanate solution drop wise slowly to the above suspension with constant stirring.

- Continue stirring for 5 hours at room temperature. (As reaction is exothermic the temperature will raise and mixture should be cooled to maintain it at room temperature).

- Add the resulting mixture drop-wise into ice cold solution of 20% acetic acid (200.0 ml) over a period of 30-45 minutes with constant stirring. After complete addition, chlorpropamide precipitates out.

- Filter the precipitate formed under vacuum, wash with 50.0 ml cold water.

- Dissolve the crude product in 75.0 ml of 5.0% sodium carbonate solution and filter immediately from the insoluble gummy material.

- Neutralize the filtrate with 20.0% ice cold acetic acid while cooling the solution to precipitate the product.

- Filter the resulting precipitate under vacuum, dry and recrystallize from benzene to give the pure product.

Physical properties:

Melting point: 130°C.

Percentage yield: 60% $^W/_W$

Category:

It is an antihypertensive agent belonging to the class of sulphonyl ureas.

It is indicated in diabetes insipidus occasionally.

References:

1. Raghu Prasad, M.; Raghuram Rao, A. Synthesis of chlorpropamide. Unpublished results.

## 15

Preparation of Aspirin

Aim: To prepare aspirin from salicylic acid.

Apparatus: Conical flask, Buchner funnel, glass rod etc.

Retrosynthetic Analysis:

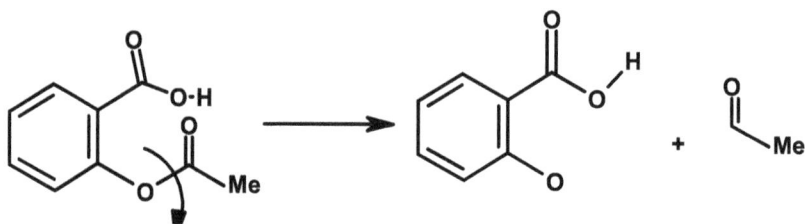

Aspirin molecule can be split into o-oxy benzoic acid and acetyl groups. By adding hydrogen to oxo functional group of o-oxy benzoic acid results in salicylic acid which can serve as one of the synthons. The acetyl group can be obtained by acylating agents like acetyl chloride or acetic anhydride. Thus these reagents can serve as one of the synthons.

Chemicals:

Chemical & Reagents	Mol. Wt.	Mol. Formula	Moles	Quantity Taken
Salicylic acid	122	$C_7H_6O_2$	0.01 M	1.22 g
Acetic Anhydride	102	$C_4H_6O_3$	0.5 M	5.04 g
Conc. Sulphuric acid	82	H_2SO_4		2-3 drops
Ethanol	46	C_2H_5OH		q.s

Principle:

The principle involved in the above reaction is O-acylation. The lone pair of electrons on the oxygen of hydroxy group of salicylic acid attacks on the carbonyl carbon of acetic anhydride to give acetyl salicylic acid and acetic acid. The concentrated sulphuric acid protonates the oxygen of acetic anhydride and increases the positivity of the carbonyl carbon and thus hastens up the reaction process acting as a catalyst.

Acetyl salicylic acid can also be prepared by acetylation with acetyl chloride. However it needs the usage of dry pyridine which is not ecofriendly in nature. Hence the above mentioned method is preferred.

Reaction:

$$H_2SO_4 \longrightarrow H^{(+)} + HSO_4^{(-)}$$

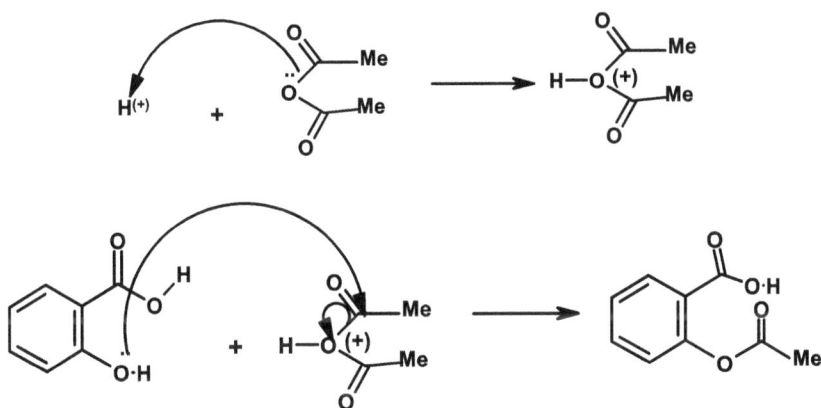

Procedure:

- Take salicylic acid (1.22 g, 0.01 M) in a clean and dry conical flask.

- Add acetic anhydride (5.0 g, 0.5 M) followed by 2-3 drops of concentrated sulphuric acid and mix well.

- Heat the reaction mixture on a water bath for 30 minutes with occasional shaking.

- Cool the reaction mixture to room temperature and pour it onto crushed ice with constant stirring.

- Filter off the acetyl salicylic acid under vacuum, dry and recrystallize using ethanol as solvent.

Physical Properties:

Melting point: 118°C.

Percentage Yield: 85% $^W/_W$

Category:

- It is an analgesic indicated in mild to moderate pain.
- It is also indicated in prophylaxis of myocardial infarction and to reduce the transient ischemic attack at lower dose level.
- It is used in the treatment of arthritic conditions.

References:

1. Gerhardt, C. *Ann.* **1853**, *87*, 149.

## 16

Preparation of Guaifenesin

Aim: To prepare guaifenesin from o-methoxy phenol.

Apparatus: Conical flask, ice bath, Buchner funnel.

Retrosynthetic Analysis:

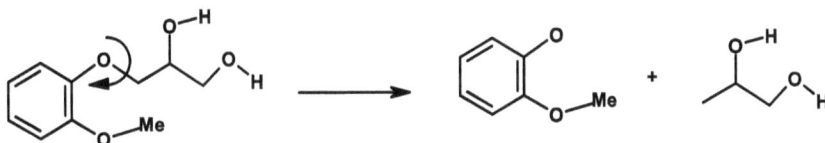

Guaifenesin can be split into o-oxy phenyl methyl ether and 1,2-dihydroxy propane fragments. Adding hydrogen to oxo function of o-oxy phenyl methyl ether, results in o-hydroxy phenyl methyl ether which serves as one of the synthons. Linking oxo group of 2-hydroxy group of 1,2-dihydroxy propane results in glycidol which serves as another synthon for the preparation of the title compound.

Chemicals:

Chemical & Reagents	Mol. Wt.	Mol. For.	Moles	Quantity Taken
o-Methoxy phenol	112	$C_6H_8O_2$	0.02 M	2.24 g
Glycidol	74	$C_3H_6O_2$	0.02 M	1.50 g
Pyridine	79	C_5H_5N	0.1 M	10.0 ml
Ethanol	46	C_2H_5OH		*q.s.*

Principle:

The Glycidol condenses with o-methoxy phenol to give guaifenesin. It is an exothermic reaction. The lone pair of electrons present on the oxygen atom of hydroxyl group of phenol attacks the positive carbon attached to oxygen and ring opening takes place to give the target compound in quantitative yields.

Reaction:

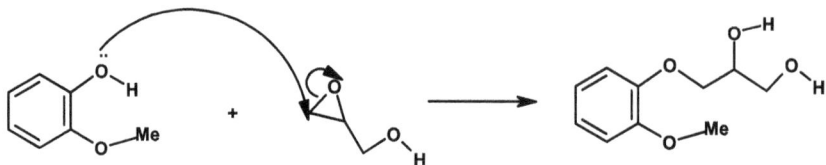

Procedure:

- Mix o-methoxy phenol (2.24 g, 0.02 M) and glycidol (1.48 g, 0.02 M) in pyridine (10.0 ml, 0.5 M) and stir it vigorously.
- Heat the reaction mixture slowly that the temperature rises to 95°C.
- As it is an exothermic reaction it is necessary to note that the temperature of reaction mixture should never cross 110°C at any point of time.
- Stir the reaction mixture for one more hour maintaining the temperature at 95°C.
- Distill off the reaction mixture at reduced pressure and collect the major fraction boiling between 175°C-180°C at 0.5 mmHg pressure.
- Crystals of guaifenesin crystallizes upon cooling the distillate.
- Filter the crystals formed, dry and recrystallize by minimum amount of ethanol to give the pure product.

Physical Properties:

Melting Point: 79°C.

Percentage yield: 62% $^W/_W$

Category:

- Used as expectorant in cough mixtures.
- It is used along with theophylline as bronchodilator.
- It is a centrally acting muscle relaxant.

References:

1. Marle *et al.*, *J. Chem. Soc.* **1912**, *101*, 305.
2. Yale *et al.*, *J. Am. Chem. Soc.* **1950**, *72*, 3710.
3. Merk, W. et al., DE 3106995.

Preparations involving Microwave Assistance

## 17

Microwave Assisted Synthesis of 1,4-Dihydropyridine

Aim: To prepare 1, 4-dihydropyridine by microwave assisted synthesis.

Apparatus: Beaker, funnel, glass rod, microwave oven.

Retrosynthetic Analysis:

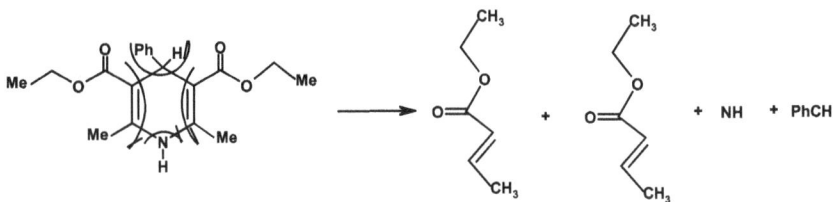

Splitting of 1,4-dihydropyridine gives 2 molecules of ethylbutenoate, amino group and benzyl group. Ethylacetoacetate can serve as a source for ethylbutenoate. Ammonia and benzaldehyde can serve as a source for amino and benzyl groups. Ethyl acetoacetate can react with ammonia to give the intermediate, ethyl ester of 3-amino-2-butenoic acid. In the presence of ammonia, methylene group can loose a proton and generate a carbanion which reacts with benzaldehyde to give a condensed product, 1-phenyl-3-carbethoxy-but-2-en-3-one. Both the products can in turn condense to yield the target compound.

Chemicals:

Chemicals & Reagents	Mol. Wt.	Mol. Formula	Moles	Quantity Taken
Benzaldehyde	106	C_7H_6O	0.01 M	1.06 g
Ethylacetoacetate	114	$C_6H_{10}O_2$	0.02 M	2.28 g (2.5 ml)
Ammonia	17	NH_3	0.01 M	0.17 g
Methanol	32	CH_3OH		15.0 ml

Principle:

A wide variety of pyridine derivatives are prepared by Hantzch method which involves condensation of a β-keto ester with an aldehyde and ammonia.

Organic reactions using microwave heating is more advantageous over conventional heating method as they provide greater reaction rates, better selectivity and thus give cleaner products as less amount of byproducts are formed. Above all it is simple and ecofriendly.

Ammonia reacts with ketonic carbonyl of ethylacetoacetate to give ethyl ester of 3-amino-2-butenoic acid. In the presence of base like ammonia, methylene group loses a proton and gives rise to a carbanion which reacts with benzaldehyde to give a condensed product, 1-phenyl-3-carbethoxy-but-2-en-3-one. This condensation is known as Knoevenagel condensation. Both this products again condense to give 2,6-dimethyl-3,5-diethylaceto-4-phenyl-1,4-dihydropyridine in quantitative yields.

Reactions:

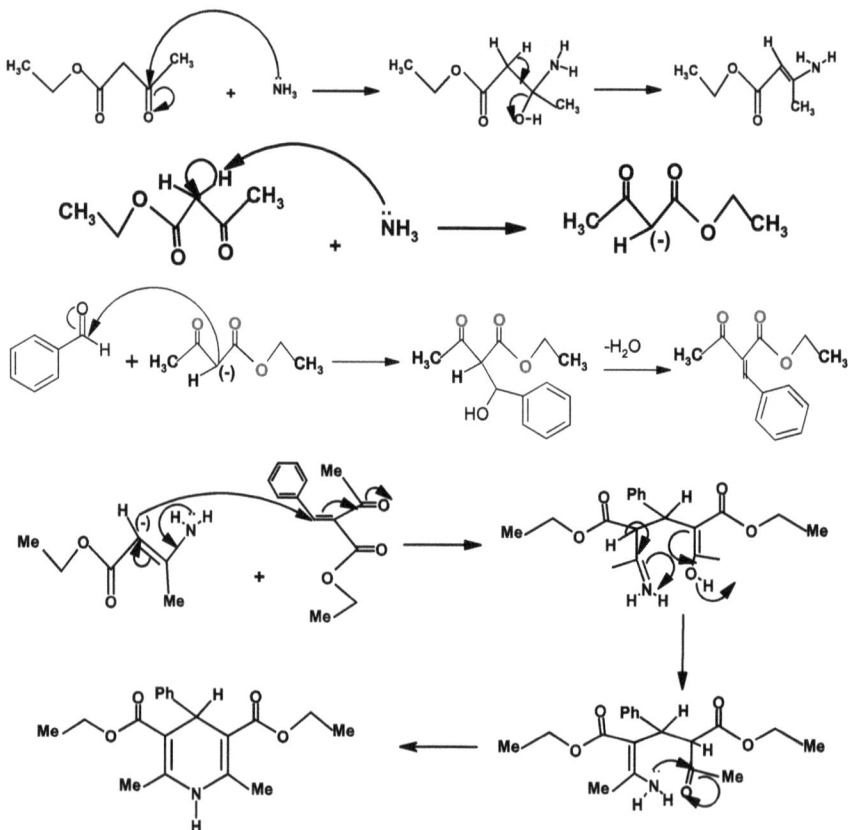

Procedure:

- Place benzaldehyde (1.06 g, 0.01 M), ethylacetoacetate (2.28 g, 2.5 ml, 0.02 M) and ammonia (0.17 g, 0.01 M) in a beaker and mix with methanol (15.0 ml).

- Irradiate the reaction mixture in a microwave oven at 480 watts for 4 minutes. (The irradiation was stopped every 30 seconds once; 2.0 ml of ammonia was added and continued).

- After the completion of the reaction, cool the contents of the beaker to room temperature and add crushed ice in which product precipitates out.

- Filter of the product obtained and dry.

Physical Properties:

Melting Point: 216°C-218°C.

Percentage yield: 85% $^W/_W$

Category:

Antihypertensive and antiarrhythmic agent.

References:

1. Hantzsch Cantwell, N.; Brown, E. V. *J. Am. Chem. Soc.* **1952**, *74*, 5967.

2. Raghu Prasad, M.; Raghuram Rao, A. Microwave assisted synthesis of dihydro pyridine. Unpublished results.

Preparations involving two step syntheses

18

Preparation of Benzoic Acid

(Synonym: phenyl formic acid)

Aim: To prepare benzoic acid from phenol and benzoyl chloride.

Apparatus: Conical flask, beaker, rubber cork, funnel, filter paper.

Retrosynthetic Analysis:

There are number of methods for the preparation of benzoic acid available in literature. The most widely used methods are oxidation of toluene by potassium permanganate and acidic hydrolysis of benzonitrile. However they involve tedious and cumbersome work up procedures. Other methods include alkaline hydrolysis of esters like phenyl benzoate to give corresponding acid and phenol respectively. It is not a versatile method as esters are in turn prepared by esterification of corresponding acids. However it is used rarely as a method of preparation of benzoic acid.

Chemicals:

Chemicals & Reagents	Mol. Wt.	Mol. Formula	Moles	Quantity Taken
Benzoyl Chloride	140.5	C_7H_5OCl	0.010 M	1.45 g (1.2 ml)
Phenol	94	C_6H_6O	0.014 M	1.285 g (1.2 ml)
Sodium Hydroxide (10 %)	40	NaOH	3.75 M	15.0 ml
Dilute sulphuric acid	98	H_2SO_4		*q.s.*
Denatured spirit	46	C_2H_5OH		*q.s.*

Principle:

Principle involved in the preparation of benzoic acid is hydrolysis of phenyl benzoate. Phenyl benzoate can be prepared by Schotten-Baumann reaction. The hydroxy (phenol) compound reacts with benzoyl chloride under basic condition to give the ester, phenyl benzoate. Alkaline phenolic hydrolysis of formed benzoate gives benzoic acid and phenol.

Reaction:

Step I

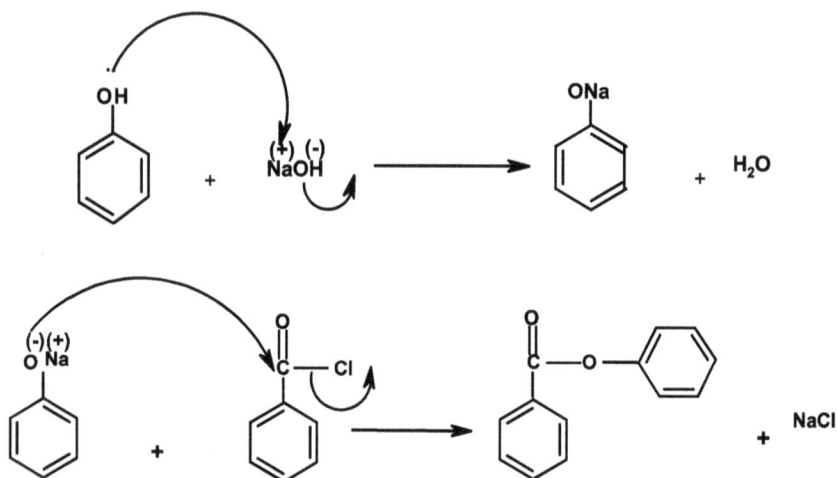

Step II

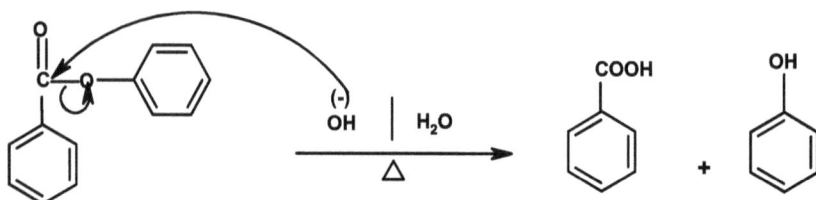

Procedure:

Preparation of phenyl benzoate from phenol

- Take phenol (1.45 g, 1.2 ml, 0.010 M) and 15 ml of 10 % NaOH solution in a conical flask and close it with a rubber cork. Shake the contents of the flask until the phenol dissolves completely.

- Add benzoyl chloride (2.285 g, 2.0 ml, 0.014 M) in small portions and shake vigorously. Phenyl benzoate forms as a fine precipitate.
- Filter the resultant precipitate at the pump, dry and recrystallize from denatured spirit.

Preparation of benzoic acid from phenyl benzoate

- Take phenyl benzoate (obtained from the first step) and 15.0 ml of 10 % NaOH in a round bottomed flask. Fit with a reflux condenser and heat the reaction mixture to reflux until the molten ester completely dissolves.
- A small amount of ester may volatilize in steam and deposits again in the condenser. Add 3.0 ml – 5.0 ml of 10 % NaOH along the sides of the condenser from the top to dislodge the crystallized ester.
- Continue heating until the solution remained clear.
- Cool the reaction mixture in ice water and neutralize with dilute H_2SO_4 until a precipitate of benzoic acid separates out.
- Filter the resultant precipitate at pump, dry in an oven and recrystalize from ethanol.

Physical Properties:

Melting Point: 120°C.

Percentage yield: 90% $^W/_W$

Category:

- Antifungal and antiinfective agent.
- Used as an antiseptic, ointments and mouth washes.
- Employed with salicylic acid as topical antifungal agent
- Used as a preservative in pharmaceutical industry.

Identification tests:

To little amount of compound, added 2. 0 ml of nitrating mixture and left aside for 2-5 minutes. The solution was poured into a beaker containing water. A faint yellow solid of m-nitro benzoic acid separates out.

References:

1. Vogel, A. I. Text book of Practical Organic Chemistry. Addison Wesley Longmann Ltd. England. 5th Ed. **1989**, PP-1274.

2. Arun, S. *Systematic lab experiments in Organic Chemistry*. New Age International (P) Ltd. Publishers, New Delhi, 2nd Ed., **2006**, PP-263, 619.

19

Preparation of 1-Amino-2-Hydroxy Napthalene HCl

Aim: To prepare 1-amino-2-hydroxy napthalene hydrochloride from aniline.

Apparatus: Reflux condenser, conical flask, beaker, funnel, water bath.

Retrosynthetic Analysis:

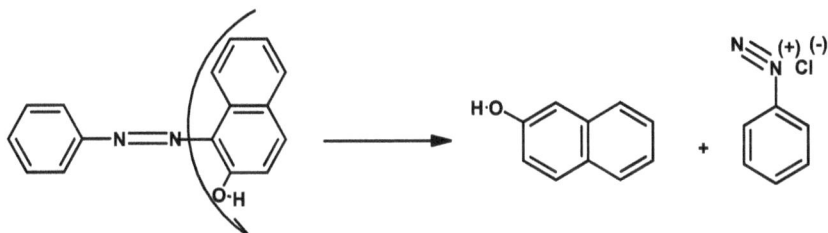

1-Amino-2-hydroxy naphthalene can be prepared by reduction of phenyl-azo-β-naphthol. Splitting of phenyl-azo-β-naphthol gives rise to β-napthol and benzene diazonium hydrochloride. Hence we can use them as synthons for its preparation. Benzene diazonium chloride can be prepared by diazotization of aniline using sodium nitrite and concentrated hydrochloric acid.

Chemicals:

Chemicals & Reagents	Mol. Wt.	Mol. Formula	Moles	Quantity taken
Aniline	93	C_6H_7N	0.013 M	1.24 g (1.22 ml)
Sodium nitrite	69	$NaNO_2$	0.015 M	1.0 g
β-Napthol	144	$C_{10}H_8O$	0.014 M	1.96 g
Tin chloride	188.7	$SnCl_2$	0.011 M	2.0 g
Concentrated HCl	36.5	HCl	0.05 M	4.0 ml
10 % Sodium hydroxide	40	NaOH	0.0025 M	10.0 ml

Principle:

Principle involved in the preparation of 1-amino 2-hydroxy napthalene involves 2 steps.

Step1: Preparation of 1-Phenyl azo β-napthol from aniline.

Diazonium salts react with phenols to give azo compounds having general formula Ar-N=N-Ar. This reaction is known as coupling reaction and the aromatic compound couple with diazonium salts containing electron releasing functional groups like -OH, -OR, NHR -NH etc. Coupling occurs in ortho or para positions with respect to activating group.

   Aniline on diazotization with sodium nitrite and HCl gives benzene diazonium chloride this then couples with β-Naphthol in the presence of alkali to give phenyl azo β-naphthol.

Step2: Preparation of 1-amino 2-hydroxy naphthalene HCl from phenyl-azo-β-naphthol.

The principle involved in this reaction is reduction. Reduction of phenyl azo β-naphthol with stannous chloride in presence of HCl gives 1-amino 2-hydroxy naphthalene HCl. Prolonged reduction first reduces azo group to give an intermediate hydrazo derivative and then breaks NH-NH linkage with formation of 2 primary amines, aniline and 1-amino-2-hydroxy naphthalene.

Reaction:

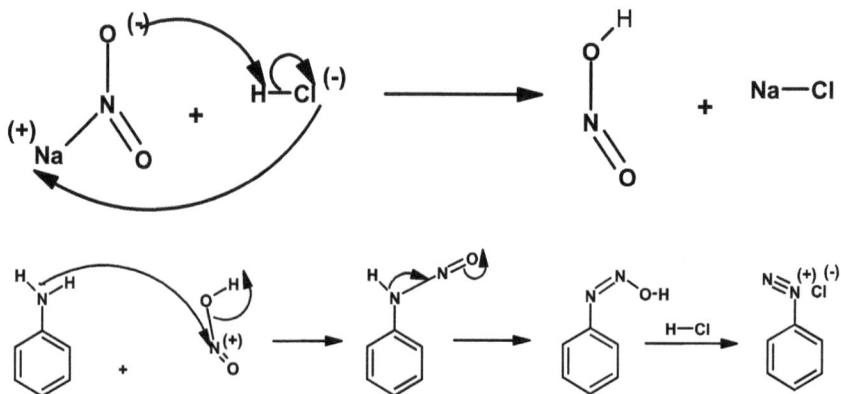

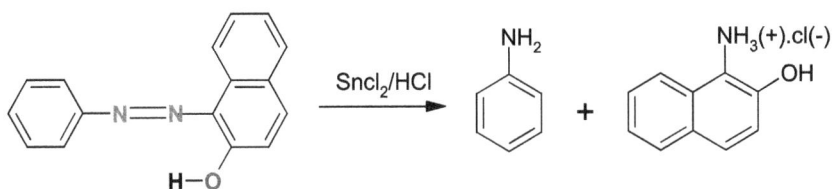

Procedure:

Step1: Preparation of phenyl azo β-Napthol from aniline.

- Add aniline (1.22 ml, 0.013 M) to a mixture of concentrated HCl (4.0 ml) and water (4.0 ml) taken in a beaker and cool it by keeping in an ice bath.

- Add an ice cold solution of sodium nitrite (1.0 g, 0.015 M) in water (5.0 ml) to the above mixture slowly with stirring.

- Prepare a solution of β-Naphthol by dissolving β-Naphthol (1.96 g, 0.014 M) in 10% NaOH (10.0 ml) in a separate beaker and cool to 5°C by immersing in an ice bath.

- Stir the naphthol solution vigorously and add diazonium solution carefully. Red crystals of phenyl azo β-Naphthol separates out.

- Allow the mixture to stand in an ice bath for about 20 minutes to aid complete precipitation.

- Filter the precipitate obtained, dry in a hot air oven and recrystallize from ethanol.

Step2: Preparation of 1-amino 2-hydroxy naphthalene from phenyl azo β-Naphthol.

- Take the phenyl azo β-Naphthol (2.0 g, 0.008 M) and ethanol (10.0 ml) in a 100 ml round bottomed flask fitted with a reflux condenser.

- Meanwhile dissolve tin chloride (2.0 g, 0.011 M) in concentrated HCl (5.0 ml) and add to the above reaction mixture.

- Heat the mixture to reflux for about 30 minutes until all the dye dissolves and rapidly gets reduced by tin chloride. The solution turns to pale green in color.

- Add the reaction mixture onto crushed ice, filter off the precipitated solid, dry and recrystallize from a minimum quantity of ethanol.

Physical Properties:

Melting Point: 150°C.

Percentage yield: 30% $^W/_W$

Category:

- Scarlet dye.
- Used in the preparation of 1, 2-napthaquinone.
- As a biological stain.

References:

1. Vogel, A. I. Text book of Practical Organic Chemistry. Addison Wesley Longmann Ltd. England. 5th Ed. **1989**, PP 949.
2. Mann, F. G.; Saunders, B. C. *Practical Organic Chemistry*, Orient Longmann Ltd. Hyderabad. 4th Ed. **1990**, pp 210-212.

## 20

Preparation of 7-Hydroxy-4-Methyl-8-Nitro Coumarin

Aim: To prepare 7-hydroxy-4-methyl-6/8-nitro-coumarin from resorcinol.

Apparatus: Beaker, conical flask, spatula, glass rod.

Retrosynthetic Analysis:

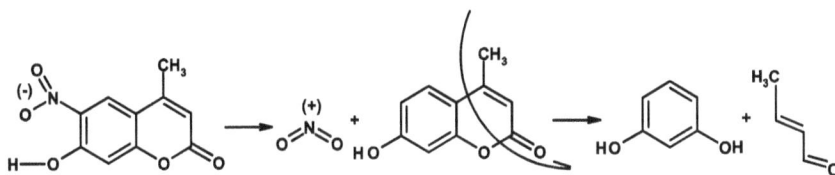

The splitting of 7-hydroxy-4-methyl-8-nitro-coumarin gives rise to the following, (a) nitro group, (b) resorcinol and (c) a four-carbon donor. The nitro group can be obtained by nitrating mixture of concentrated H_2SO_4 and concentrated HNO_3. Ethylacetoacetate serves as a four carbon donor. By combining ethylacetoacetate and resorcinol we can get coumarin which can be nitrated using a nitrating mixture to yield the target compound.

Chemicals:

Chemicals & Reagents	Mol. Wt.	Mol. Formula	Moles	Quantity Taken
Resorcinol	110	$C_6H_6O_2$	0.009 M	0.95 g
Ethylacetoacetate	114	$C_6H_{10}O_2$	0.009 M	1.0 g (1.1 ml)
Concentrated nitric acid	63	HNO_3		0.33 ml
Concentrated sulphuric acid	98	H_2SO_4		3.75 + 0.661 + 6.751
Dilute hydrochloric acid	36.5	HCl		*q.s.*
Aqous sodium hydroxide	40	NaOH		*q.s.*

Principle:

It involves 2 steps

1. Preparation of 7-hydroxy-4-methyl coumarin from resorcinol.

2. Preparation of 8-nitro-7-hydroxy-4-methyl coumarin from the above product.

Preparation of 7-hydroxy-4-methyl coumarin from resorcinol.

The principle involved is Pechmann condensation. Resorcinol condenses with β-keto esters in presence of a dehydrating agent, concentrated H_2SO_4 or polyphosphoric acid to give coumarin.

β-keto ester like ethylacetoacetate exhibits keto enol tautomerism. The enolic form reacts with phenol, resorcinol in the presence of concentrated H_2SO_4 to give substituted coumarin by loosing a molecule of water and ethanol.

Alternatively, the reaction can be carried out by microwave assistance which completes within 30 seconds. Instead of concentrated H_2SO_4, p-toluene sulphonic acid is used as a dehydrocyclizing agent.

Pechmann reaction is mildly exothermic but not spontaneous. The short bust of energy given by microwave irradiation initiates the reaction and proceeds to completion.

Preparation of 7-hydroxy-4-methyl-8-nitro-coumarin from the above product.

The principle involved in step-2 is nitration. Aromatic hydrocarbons when treated with nitric acid in the presence of H_2SO_4 undergo direct nitration to give respective nitro derivative. Here nitration of 7-Hydroxy-4-methyl-coumarin gives both 6 and 8 nitro derivatives.

Reaction:

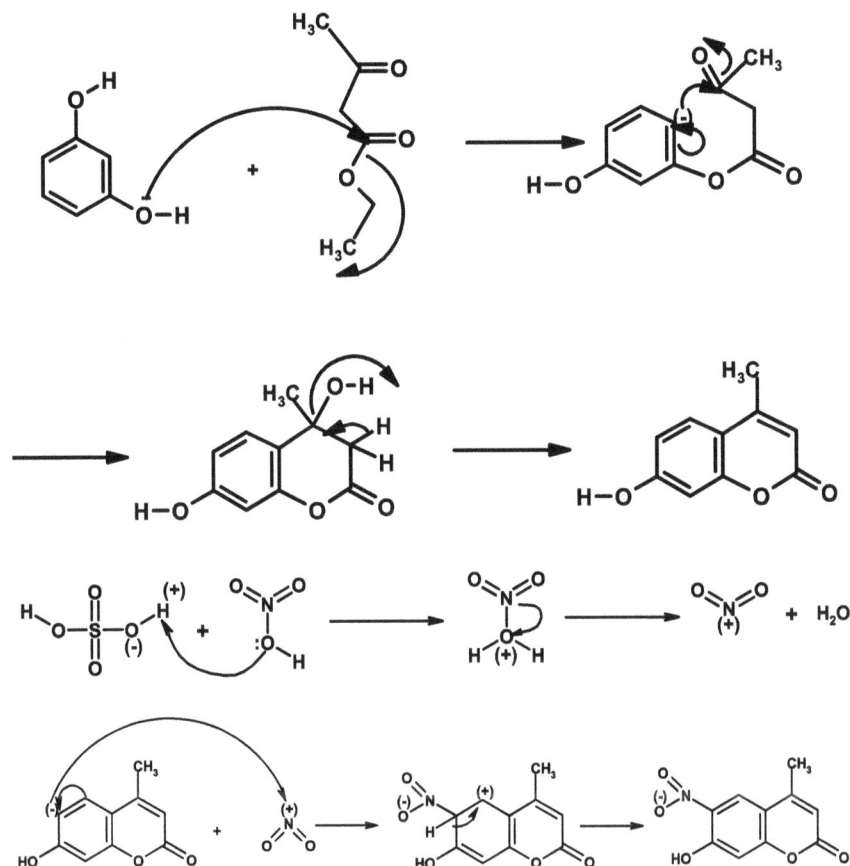

Procedure:

Step-1: Preparation of 7-hydroxy-4-methyl coumarin from resorcinol.

- Take concentrated sulphuric acid (3.75 ml) in a conical flask and stirr mechanically with external cooling such that the temperature does not rise above 5°C.

- Meanwhile, take powdered resorcinol (0.95 g, 0.009 M) in a separate beaker, add ethylacetoacetate (1.1 ml, 0.009 M) and stir until a complete dissolution is affected.

- Add resulting solution slowly to concentrated H_2SO_4 with stirring by maintaining the temperature of the mixture at $0°C$-$5°C$.

- Continue stirring for about 30 minutes and pour the reaction mixture onto 50.0 g of crushed ice.

- Separate the precipitate formed by filtration at the pump and dry in a hot air oven.

- Purify the product by dissolving in 10 % cold aqueous NaOH and then reprecipitating with dil. HCl.

- Rrecrystallize the product from ethanol for further purification, if necessary.

Alternate microwave assisted method.

- Take a mixture of resorcinol (0.95 g, 0.009 M), ethylacetoacetate (1.1 g, 0.009 M) and p-toluene sulphonic acid (0.09 g, 0.0005 M) in a 25.0 ml conical flask.

- Irradiate the mixture at 400 W for 30 seconds in a domestic microwave oven. After irradiation, temperature raises to $85°C$.

- Remove the reaction mixture from the oven, cool it to room temperature and add crushed ice.

- Filter off the precipitated coumarin, dry and purify it by recrystallization with methanol.

Physical Properties:

Melting Point: $185°C$.

Percentage yield: 90% $^W/_W$ (95% in microwave method).

Step-2: Preparation of 8-nitro-7-hydroxy-4-methyl coumarin

- Dissolve the 7-hydroxy-4-methyl coumarin, the product of 1st step in concentrated H_2SO_4 (6.75 ml) contained in a 100 ml conical flask and cool in an ice bath maintained, below $10°C$.

- Add solution of a mixture of concentrated H_2SO_4 (0.66 ml) and concentrated nitric acid (HNO_3) (0.33 ml) dropwise to the above solution maintaining the temperature below $10°C$.

- Allow the reaction mixture to stand in an ice bath for about 30 minutes.

- Pour the contents of the flask onto crushed ice and filter off the precipitated nitro derivatives of 7-hydroxy-4-methyl coumarin.

- Boil the product with ethanol (15.0 ml) and filter while hot. The residue obtained is 6-nitro-7-hydroxy-4-methyl coumarin which is purified by recrystallization with benzene.

- Collect the filtrate and evaporate to half of its volume and cool in an ice bath where 8-nitro-7-hydroxy-4-methyl coumarin is obtained as yellow crystalline solid which may be further purified by recrystallization from ethanol.

Physical Properties:

Melting Point: 262°C (6-nitro derivative), 256°C (8-nitro derivative).

Percentage yield: 53% $^W/_W$

Category:

Anticoagulant, antitumor, vasodilator, antibacterial agent.

Finds its use as an antispasmodic and chloerectic.

References:

1. Arun, S. *Systematic Lab Experiments in Organic Chemistry*. New Age International (P) Ltd. Publishers, New Delhi, 2nd Ed., **2006**, PP 676.

2. Mann, F. G.; Saunders, B. C. *Practical Organic Chemistry*, Orient Longmann Ltd. Hyderabad. 4th Ed. **1990**, pp 307.

3. Manhas, M. S.; Ganguly, S. N.; Mukherjee S.; Amit,. K. J.; Ajay, K. B. *Tetrahedron Lett.* **2006**, *47*, 2423.

21

Preparation of Antipyrine

(Synonyms: Analgesine, Anodynine, Phenylone, Sedatine, Parodyne)

Aim: To prepare antipyrine from ethylacetoacetate and phenyl hydrazine.

Apparatus: Evaporating dish, water bath, glass rod, Buchner funnel, Buchner assembly, reflux condenser.

Retrosynthetic Analysis:

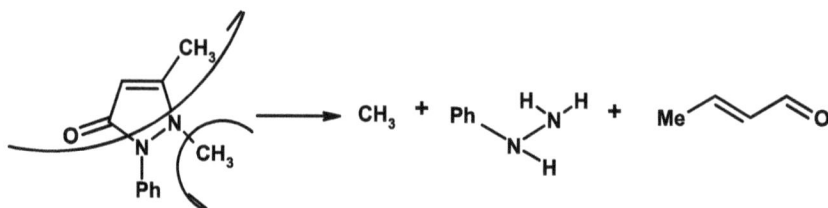

Antipyrine moiety can split into phenyl hydrazine, methyl and a four carbon donor. Dimethyl sulphate or methyl iodide can serve as a methyl donor and ethylacetoacetate can serve as a 4 carbon donor. Thus phenyl hydrazine, ethylacetoacetate and dimethyl sulphate can be used as synthons for the preparation of antipyrine.

Chemicals:

Chemicals & Reagents	Mol. Wt.	Mol. Formula	Moles	Quantity Taken
Phenyl hydrazine	108	$C_6H_8N_2$	0.04 M	4.0 g
Ethylacetoacetate	114	$C_6H_{10}O_2$	0.04 M	5.0 g
Dimethyl suphate	126.3	$C_2H_6SO_4$	0.03 M	3.6 g (2.85 ml)
3-Methyl-1-phenyl pyrazolone	174	$C_{10}H_{10}N_2O$	0.025 M	4.35 g
Sodium hydroxide	40	NaOH	0.025 M	1.0g
Diethyl ether	74	$C_4H_{10}O$		10.0 ml

Principle:

Preparation of antipyrine involves 2 steps.

Step-I: Preparation of 3-methyl-1-phenyl pyrazol-5-one.

This is an example for condensation reaction. The lone pair of electrons on nitrogen of amino group attacks on the carbonyl carbon of the ethyl acetoacetate portion of ethyl acetoacetate to give an intermediate 2-phenyl-hydrazino-2-hydroxy-ethyl butyrate. It undergoes cyclization by removal of ethanol to give the target compound in quantitative yields.

Step-II: Preparation of 2, 3-dimethyl-1-phenyl pyrazol-5-one (Antipyrine).

In this step methylation of secondary nitrogen present in the pyrazolone moiety is carried out by using a methylating agent, dimethylsulphate under strongly basic condition. The lone pair of electrons from the secondary nitrogen in the pyrazolone ring attacks on the positive methyl portion of dimethylsulphate to give the target compound, antipyrine.

Reaction:

Step-I

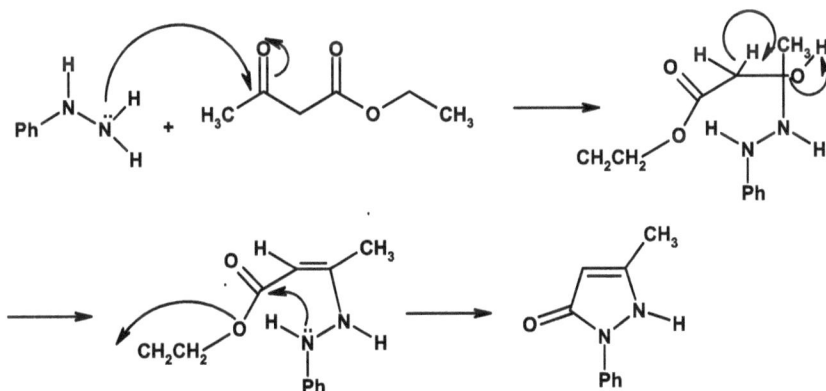

Step-II

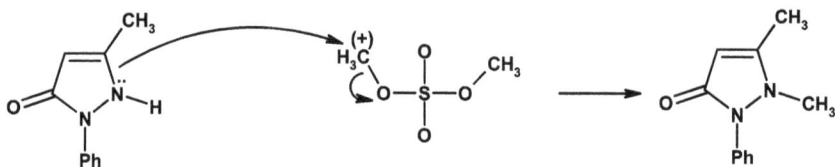

Procedure:

Step-I: Preparation of 3-methyl-1-phenyl pyrazol-5-one.

- Mix redistilled ethylacetoacetate (5.0 g, 4.9 ml, 0.04 M) and phenyl hydrazine (4.0 g, 3.65 ml, 0.04 M) in an evaporating dish and heat it on a boiling water bath kept in an efficient fuming cupboard for about 2 hours by stirring with a glass rod frequently.

- After cooling to room temperature, to the resulting heavy reddish syrup add ether (10.0 ml) and stir vigoursly.

- The syrup insoluble in ether readily solidifies within 15 minutes.

- Filter off the precipitate obtained and wash with ether several times to remove colored impurities if any.

- Dry the compound obtained in a hot air oven and purify by recrystallization using equal volumes of ethanol and water.

Physical Properties:

Melting Point: 127°C.

Percentage yield: 70% $^W/_W$

Step-II: Preparation of 2, 3-dimethyl-1-phenyl pyrazol-5-one (Antipyrine)

- Take an accurately weighed amount of 3-methyl-1-phenyl pyrazol-5-one (4.35 g, 0.025 M), sodium hydroxide (1.0 g, 0.025 M) [dissolved minimum possible quantity of water] and methanol (5.0 ml) in a 3-necked flask fitted with a reflux condenser and a dropping funnel.

- Warm the reaction mixture on a water bath and add dimethyl sulphate[*] (3.6 g, 2.7 ml, 0.03 M) through a dropping funnel and continue heating to reflux for an hour.

*Dimethyl sulphate (DMS) is a highly corrosive and alkylating reagent. It should be handled carefully and all the procedures involving DMS, should be conducted in an efficient fume hood.

- Distill the excess of methanol, add hot water to the resulting residue and stir with the help of a glass rod.

- Filter off the above mixture to get rid of the impurities and extract the filtrate with toluene.

- Separate the benzene layer, dry it over anhydrous sodium sulphate and distill off to get the crude product which is purified by recrystallization by benzene or petroleum ether.

Physical Properties:

Melting Point: 113°C.

Percentage yield: 60% $^W/_W$

Category:

- Analgesic and anti-inflammatory agent.

References:

1. Mueller, A. *Monatsh.* **1958**, *89*, 23.

2. *Hagers Handb. Pharm. Praxis.* Springer Verlag, Berlin, **1977**, *6*, pp 571.

## 22

Preparation of Paracetamol and Phenacetin

Aim: To prepare paracetamol and phenacetin from p-amino phenol.

Apparatus: Conical flask, water bath, Buchner filtration assembly, round bottomed flask, reflux condenser.

Retrosynthetic Analysis:

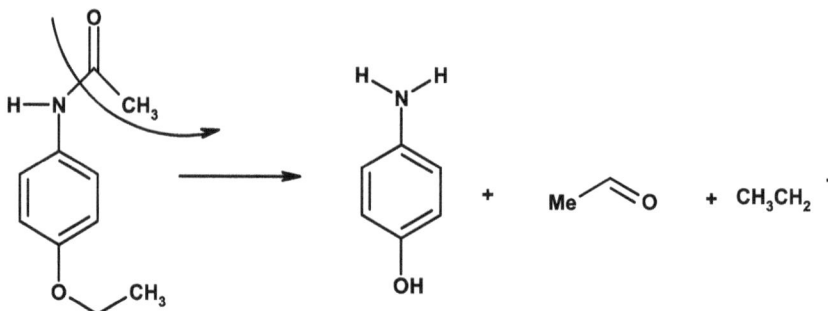

The phenacetin molecule can split into p-amino phenol, acetyl group and ethyl moiety. Acetic anhydride and ethyl iodide can serve as acetyl group and ethyl group donors. Hence one can use p-amino phenol, acetic anhydride and ethyl iodide as synthons for the preparation of paracetamol and phenacetin respectively.

Chemicals:

Chemicals & Reagents	Mol. Wt.	Mol. Formula	Moles	Quantity Taken
p-Amino phenol	109	C_6H_7NO	0.010 M	1.1 g
Acetic anhydride	102	$C_4H_6O_3$	0.012 M	1.2 g
Sodium	23	Na	0.007 M	0.161 g

Table *Contd...*

Chemicals & Reagents	Mol. Wt.	Mol. Formula	Moles	Quantity Taken
p-Hydroxy acetanilide	151	C$_8$H$_9$NO$_2$	0.007 M	1.0 g
Ethyl iodide	150	C$_2$H$_5$I	0.01 M	1.5 g (0.8 ml)
Absolute ethanol	46	C$_2$H$_5$OH		q.s

Reaction:

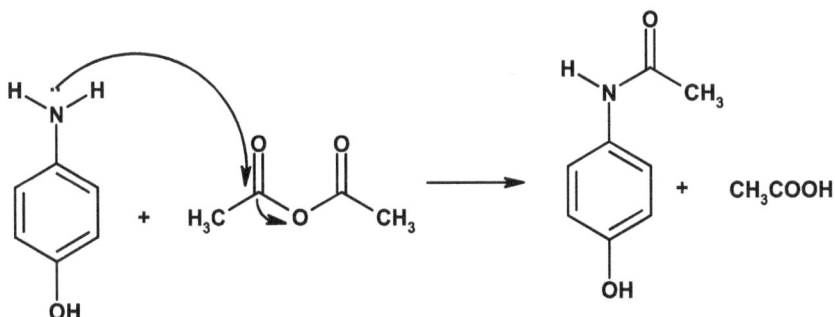

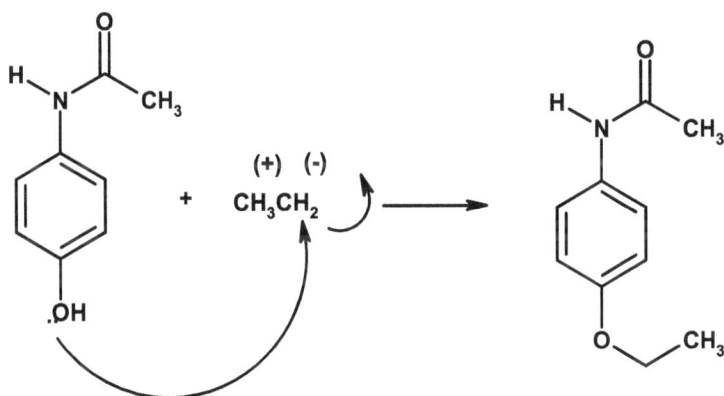

Procedure:

Preparation of paracetamol

- Suspend an accurately weighed amount of p-amino phenol (1.1 g, 0.01 M) in around 5.0 ml of water.
- Add acetic anhydride (1.2 g, 0.012 M) to the above mixture, stir vigorously and warm on a water bath until dissolution is complete (10-20 minutes).

- Cool the reaction mixture to room temperature and filter the precipitate formed, dry and purify by recrystallization using alcohol. Decolorize by charcoal treatment.

Physical Properties:

Melting Point: 169°C.

Percentage yield: 80% $^W/_W$

Category:

- Antipyretic and analgesic.
- Used in various arthritic and rheumatic conditions involving musculoskeletal pains.
- It also relives pain due to headache, dysmenorrhea, myalgia and neuralgia.
- Also finds its use as photographic chemicals and manufacturing azodyes.

Preparation of phenacetin:

- Place an accurately weighed amount of sodium (0.161 g, 0.007 M) in a 50 ml round bottomed flask and dissolve in 5.0 ml of absolute ethanol.
- To the above solution add p-hydroxy acetanilide (1.0 g, 0.007 M) followed by ethyl iodide (1.5 g, 0.8 ml, 0.01 M) along the sides of the container and heat to reflux under calcium guard tube for an hour.
- Add around 10.0 ml of water through the condenser at such a rate that the crystalline product does not separate.
- Then cool the flask by keeping on an ice bath, the crude precipitate separates. Filter the product obtained, wash with little cold water, dry in a hot air oven and purify by recrystallization using ethanol.

Physical Properties:

Melting Point: 137°C.

Percentage yield: 70% $^W/_W$

Category:

- Analgesic and antipyretic agent.
- Earlier it was the component of APC tablets but is discontinued in the early eighties as it was anticipated to be carcinogenic.

References:

1. Lumerie et al., *Bull. Soc. Chim. France.* **1905**, *33*, 785.
2. Fierz-David, K. *Helv. Chim. Acta.* **1939**, *22*, 94.
3. *Beilstein*, Vol XIII, 461.

## 23

Preparation of Propranolol

Aim: To prepare propranolol from 1-naphthol.

Apparatus: Round bottomed flask, separating funnel, magnetic stirrer, reflux condenser.

Retrosynthetic Analysis:

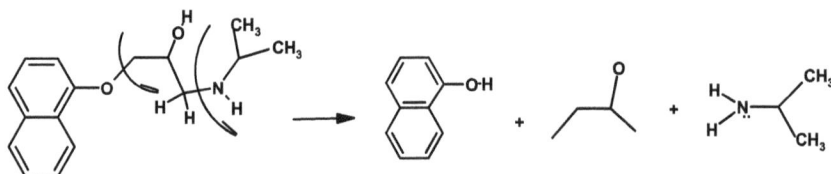

Propranolol can split into α-naphthol, isopropylamine and a propanol linkage. Epichlorohydrin can serve as a synthon for propanol linkage. Hence α-naphthol, epichlorohydrin and isopropyl amine can be used for the synthesis of the target compound.

Chemicals:

Chemicals & Reagents	Mol. Wt.	Mol. Formula	Moles	Quantity Taken
1-Naphthol	144	$C_{10}H_8O$	0.0085 M	1.25 g
Potassium hydroxide	56	KOH	0.009 M	0.50 g
Epichlorohydrin	90.5	C_3H_5OCl	0.049 M	4.0 ml
Isopropylamine	59.0	C_3H_9N	0.047 M	2.78 g (4.0 ml)
Hydrochloric acid	36.5	HCl		q.s.
Diethyl ether	74	$C_4H_{10}O$		q.s.
Ethanol	46	C_2H_5OH		q.s.

Principle:

Propranolol is a non selective β-adrenergic antagonist. It has equal affinity for both β_1 and β_2 adrenergic receptors. It is reported to have membrane stabilizing properties and the drug is lipophilic in nature. It is used in the treatment of hypertension to improve tolerance to exercise in patients with angina pectoris. It has been administered for the prevention of reinfraction in patients who have suffered from acute myocardial infraction. It is also used in cardiac arrhythmias. It is often effective in supraventricular tachyarrhythmias.

The 1-naphthol was treated with epichlorohydrin under basic conditions to get the corresponding 2 isomeric cyclopropyloxy methyl-naphthyl ethers. However it was not purified but removed in the next step by ether extraction.

The reaction should be carried out in efficient fuming cupboard as the epichlorohydrin used is highly carcinogenic in nature.

Reaction:

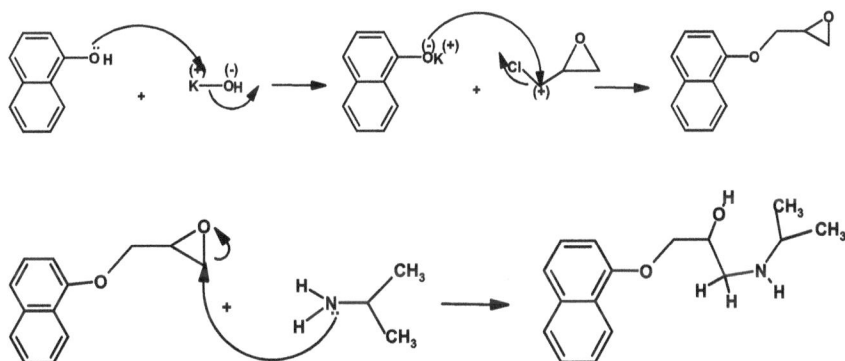

Procedure:

Preparation of 1-cyclopropyloxy methyl-naphthol ether:

- Place an accurately weighed amount of 1-naphthol (1.25 g, 0.0085M) and potassium hydroxide (0.5 g, 0.009 M) in a round bottomed flask and dissolve in 10.0 ml of ethanol-water mixture (9:1).

- Add the epichlorohydrin (4.0 ml, 0.049 M) to the above reaction mixture dropwise and stir for 48 hours.

- Remove ethanol by vacuum distillation and extract the aqueous layer with 10.0 ml portions of ether successively for 3 times.

- Combine the extracts, dry over anhydrous sodium sulphate and distill off to get the product as crude brownish colored oil.

Preparation of propranolol:

- Transfer the crude oil (0.2 g, 0.01 M) obtained in the above reaction into a round bottomed flask fitted with a stopper and dissolved in 5.0 ml of methanol.

- To the above solution, add isopropylamine (2.78 g, 4.0 ml, 0.047 M) and stir the reaction mixture at 40°C for 2 hours.

- Remove the solvent by vacuum distillation and to the residue add 30.0 ml of 2.0 M of HCl.

- Extract the resultant solution with ether and basify the aqueous extract with 32.0 ml of 2.0 M sodium hydroxide by simultaneous cooling to 0°C.

- Filter off the precipitate obtained, dry in vacuum desiccators and purify by recrystallization using petroleum ether(40-60).

Physical Properties:

Melting Point: 95°C-96°C.

Percentage yield: 67% $^W/_W$

Category:

- β-Adrenergic antagonist.

References:

1. Kaiser, C.; Jen, T.; Garvey, E.; Bowen, W. D. *J. Med. Chem.* **1997**, *20*, 687.

2. Crowther, A. F.; Smith, L. H. *J. Med. Chem.* **1968**, *11*, 1009.

3. Jaggers, S. E.; Jones, G. *J. Med. Chem.* **1978**, *21*, 182.

24

Preparation of Phenytoin

Aim: To prepare phenytoin by green chemistry technique.

Apparatus: Round bottomed flask, condenser, beaker.

Retrosynthetic Analysis:

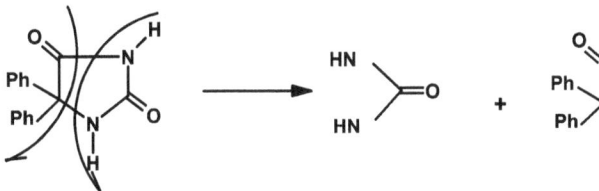

One can split phenytoin moiety into urea and 1,1-diphenyl acetaldehyde. Though urea can be used as one of the synthons, 1,1-diphenyl acetaldehyde cannot react with urea. Thus there is a need of another synthon which can react with urea and then it should be converted by further reaction or rearrangement. In this view benzil can be considered as one of the synthons which can react with urea to give the dehydrocyclized intermediate. This intermediate can undergo 1, 2-aryl shift to give the target compound.

Chemicals:

Chemicals & Reagents	Mol. Wt.	Mol. Formula	Moles	Quantity Taken
Benzoin	212	$C_{14}H_{12}O_2$	0.04 M	8.48 g
Potassium permanganate	158	$KMnO_4$	0.4 M	63.2 g
Benzil	210	$C_{14}H_{10}O_2$	0.025 M	5.3 g
Urea	60	CH_4N_2O	0.05 M	3.0 g
Sodium hydroxide (30 %)	40	NaOH		15.0 ml
Ethanol	46	C_2H_6O		75.0 ml
Dichloro methane	85	CH_2Cl_2		50.0 ml

Principle:

Preparation of phenytoin involves conversion of benzoin to benzil followed by treatment with urea to give phenytoin.

In conventional synthesis, oxidation of α-hydroxy ketone, benzoin was carried out by treating with concentrated nitric acid under reflux condition to get benzil in quantitative yields. However there will be evolution of nitrogen dioxide which is not ecofriendly in nature. Thus an alternate method involving milder oxidizing agent, like potassium permanganate could be used. Though it is a bit lengthy process but is a convenient reaction utilizing environmentally benign reagent, potassium permanganate. It gives the product in purest form than that obtained by nitric acid method.

Urea is a structural unit of 5, 5-diphenyl hydantoin and thus used as one of the synthons along with benzil. This is an excellent example of base catalyzed benzilic acid rearrangement.

It involves the nucleophilic attack on carbonyl carbons of benzil to give corresponding oxy anion which undergoes 1, 2-aryl shift to give phenytoin as shown in the reaction mentioned below.

Reaction:

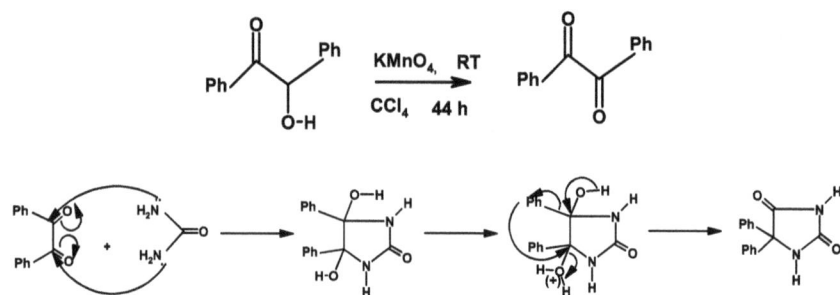

Procedure:

Step-I: Preparation of benzil from benzoin.

*   Suspend the benzoin (8.48 g, 0.04 M) and potassium permanganate (63.2 g, 0.4 M) in 100.0 ml of dichloromethane taken in a round bottomed flask.
*   Stir the reaction mixture at room temperature for 44 hours.

- After the completion of the reaction, filter off the manganese dioxide (MnO_2), dry the filtrate with magnesium sulphate ($MgSO_4$) and distill of the solvent under vacuum.

- Recrystallize the residue with carbon tetrachloride.

Physical properties:

Melting Point: 96°C.

Percentage yield: 70% $^W/_W$

Step-II: Preparation of phenytoin form benzil.

- Place benzil (5.3 g, 0.025 M), urea (3.0 g, 0.05 M), 15.0 ml of 30 % sodium hydroxide solution and ethanol (75.0 ml) in a round bottomed flask fitted with a reflux condenser and heat under reflux for 2.0 hours.

- Cool the reaction mixture to room temperature, pour onto crushed ice (125.0 g), mix thoroughly and allow it to stand for 15 minutes to remove insoluble byproducts.

- Render the filtrate strongly acidic with concentrated hydrochloric acid while cooling in an ice bath.

- Filter the precipitate obtained, dry in an oven and purify by recrystallization with ethanol.

Physical properties:

- Melting Point: 297°C-298°C.
- Percentage yield: 90% $^W/_W$

Category:

- An effective anticonvulsant used to manage generalized tonic-clonic, complex partial and simple partial seizures in the treatment of grandmal type of epilepsy.

- Used to control and manage seizures during neurosurgery.

- Useful in the treatment of paroxysmal atrial tachycardia, ventricular tachycardia and digitalis induced cardiac arrhythmias.

References:

1. Mara, E. F. B.; Hugo, T. S. B.; Marciana, P. U.; Carla, C. C.; Marcelo, S. *J. Braz. Chem. Soc.* **2008**, *19*, 909.

2. Biltz, H. *Ber.* **1908**, *41*, 1391.

3. Biltz, *Ber.* **1911**, *44*, 411.

4. Sikdar, J.; Gosh, T. N. *J. Indian Chem. Soc.* **1948**, *25*, 109.

## 25

Preparation of 6-Methyl Uracil

Aim: To prepare 6-methyl uracil from thiourea.

Apparatus: Round bottom flask, conical flask, beaker, water bath, heating mantle.

Retrosynthetic Analysis:

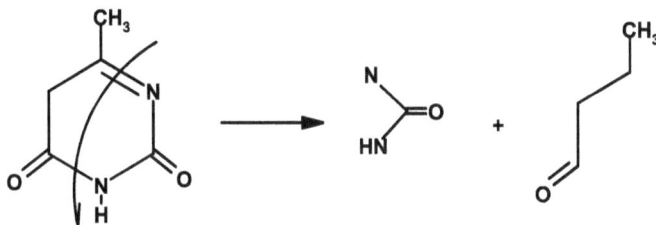

The 6-methyl uracil can be split into urea and a four carbon donor. As urea is not that reactive thiourea can be used in its place as one of the synthons. A four carbon donor like ethylacetoacetate can serve as another synthon which can react with thiourea under basic conditions to 6-methyl-2-thio uracil. This can be oxidized to 6-methyl uracil using chloroacetic acid.

Chemicals:

Chemicals & Reagents	Mol. Wt.	Mol. Formula	Moles	Quantity Taken
Thiourea	76	CH_4N_2S	0.036 M	2.8 g
Ethylacetoacetate	114	$C_6H_{10}O_2$	0.074 M	8.45 g (7.4 ml)
Potassium Hydroxide	56	KOH	0.057 M	3.2 g
Chloroacetic acid	94.5	$C_2H_3O_2Cl$	0.053 M	5.0 g
Ethanol	46	$C_2H_5O_H$		10.0 ml

Reaction:

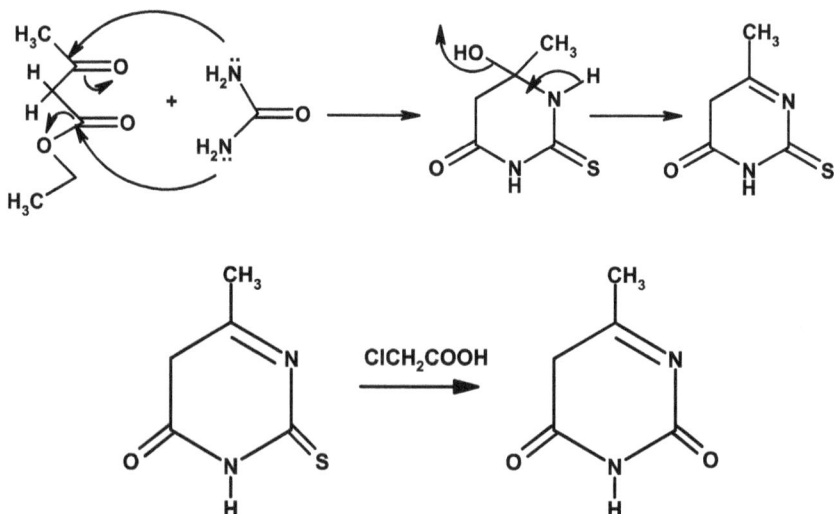

Principle:

It involves following steps

(a) Preparation of 6-methyl-4-oxo-1,2,3,4 tetrahydro-2-thiopyridine from thiourea.

(b) Preparation of 6-methyl uracil from 6-methyl-4-oxo-1,2,3,4-tetrahydro-2-thio pyridine.

The principle involves is condensation of thiourea with ethylacetoacetate in the presence of a strong base like, potassium hydroxide solution and ethanol resulting in loss of water molecule and ethanol to form required product. The principle involved in step-2 is desulfuration to produce 6-methyl uracil.

Procedure:

Step-1: Preparation of 6-methyl-4-oxo-1,2,3,4-tetrahydro-2-thiopyridine from thiourea.

- Place thiourea (2.8 g, 0.036 M) and ethylacetoacetate (7.4 ml, 0.074 M) in a 100 ml round bottom flask and dissolve in ethanol (10.0 ml).

- Add solution of potassium hydroxide (3.2 g in 4.0 ml of water, 0.058 M) to the above reaction mixture with constant stirring.
- Acidify the resultant solution with hydrochloric acid (10.0 ml concentrated HCl in 5.0 ml water).
- Cool the reaction mixture by keeping in a refrigerator in order to get a solid product.
- Separate the precipitate obtained by vacuum filtration, dry and recrystallize from ethanol.

Step-2: Preparation of 6-methyl uracil.
- Take a solution of chloroacetic acid (5.0 g, 0.053 M) in water (7.0 ml) in a 100 ml round bottom flask fitted with reflux condenser.
- Add 6-methyl-4-oxo-1,2,3,4-tetrahydro-2-thiopyridine (4.5 g, 0.05 M) to the above solution and heat under reflux on a water bath for 3 hours.
- Filter the hot solution to eliminate undissolved solid particles.
- Allow the filtrate to crystallize, separate by vacuum filtration, dry and recrystallize from ethanol.

Physical properties:
- Melting point: 246°C.
- Percentage yield: 63% $^W/_W$

Category:
- Anticancer agent.
- Used as a reagent for the synthesis of enzyme inhibitors and antiviral agents.
- Used as a reagent in the synthesis of cardiovascular agents.
- 6-Methylthiouracil acts as an antithyroid agent.

References:
1. Donleavy, J. J.; Kise, M. A. *Org. Syn.* **1943**, *Coll. Vol. II*, 422.

Preparations based on Green Chemistry

26

Preparation of 4-Bromoacetanilide (Green Chemistry)

Aim: To prepare 4-bromo acetanilide from aniline using green chemistry methods.

Apparatus: Round bottomed flask, reflux condenser, conical flasks, buchner funnel etc.

Retrosynthetic Analysis:

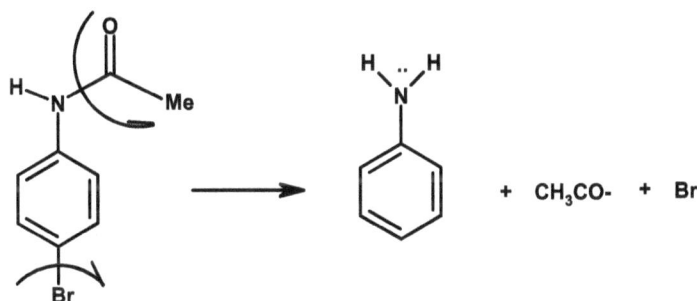

The p-Bromo acetanilide can be split into aniline, acetyl group and bromo group. Acetic anhydride can be used as acetyl group donor. However the use of acetic anhydride is restricted. Alternatively acetic acid and zinc in combination can be used as synthons to react with aniline as another synthon to yield acetanilide. Bromination can be carried out using bromine solution directly but it is not ecofriendly. Thus we can use cerric ammonium nitrate and potassium bromide in combination which generates bromonium ions and brominates acetanilide to give the target compound.

Chemicals:

Chemicals & Reagents	Mol. Wt.	Mol. For.	Moles	Quantity Taken
Aniline	93	C_6H_7N	0.1 M	9.3 g (10.0 ml)
Acetic acid	48	$C_2H_4O_2$		30.0 ml
Zinc dust	65	Zn	0.01 M	0.65 g
Acetanilide	135	C_8H_9NO	0.01 M	1.35 g
Cerric ammonium nitrate	254	$Ce(NH_3)_4NO_2$	0.02 M	6.0 g
Potassium bromide	119	KBr	0.01 M	1.19 g
Ethanol	46	C_2H_6O		15.0 ml

Principle:

In general the acylating agent used for N-acetylation is acetic anhydride. However it is restricted due to its misuse for preparation of narcotics. Thus an alternate reagent, acetic acid and zinc dust has been used. Zinc and acetic acid reacts together to form a complex which in turn acts as acylating agent for N-acylation as shown in the reaction mentioned below.

Bromination of acetanilide is carried out using bromine but it is highly corrosive. Thus an alternate environmentally benign brominating agent has been used. Potassium bromide reacts with cerric ammonium nitrate in aqueous medium give rise to bromine free radical which again by combining with cerric ion gives rise to bromonium ion which act as a brominating agent.

The electrophile, bromonium ion attacks on acetanilide at para position as the acetylamino group present on the benzene is moderately activating and o/p directing.

Reaction:

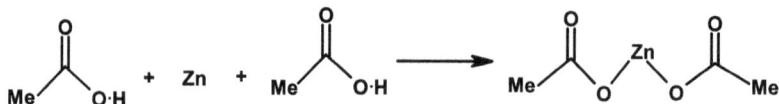

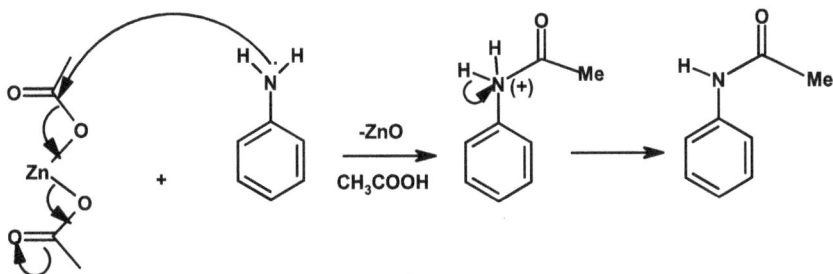

Procedure:

Step 1: Preparation of acetanilide.

- Place a mixture of aniline (10.0 ml, 9.3 g, 0.1 M), zinc (0.65 g, 0.01 M) and acetic acid (30.0 ml) in a round bottomed flask fitted with a reflux condenser.
- Heat the mixture to reflux for 2 hours.
- Cool the contents to room temperature and pour onto ice cold water (100.0 ml) slowly with vigorous stirring. Fine shining crystals of acetanilide separates out.
- After 15 minutes, filter at the pump, dry and purify by recrystallization with boiling water.

Physical Properties:

Percentage yield: 90.0% $^W/_W$

Melting Point: 114°C.

Step 2: Preparation of 4-bromo acetanilide.

- Dissolve acetanilide (1.35 g, 0.01 M) in ethanol (15.0 ml).
- Then dissolve potassium bromide (1.19 g 0.01 M) and cerric ammonium nitrate (6.0 g, 0.001 M) in water (15.0 ml).
- Add the latter solution dropwise to the former along with stirring.
- Continue stirring for 10 minutes at room temperature. After the completion of addition, white crystals separate out).
- Pour the solution onto crushed ice, filter off the precipitate formed and dry.

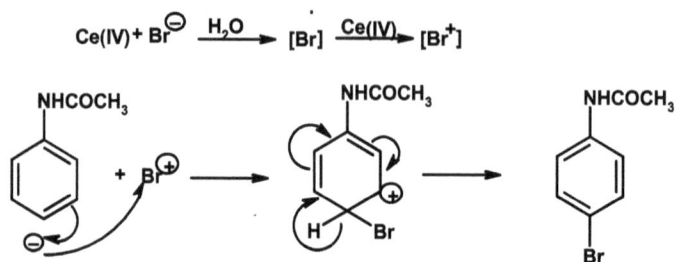

Physical Properties:

- Percentage yield: 80% $^W/_W$;   Melting Point: 165°C.

Category:

- Analgesic and anti-inflammatory.

References:

1. Meshram, S. IICT, Hyderabad, Private communication.
2. Schatz, P. F. *J. Chem. Edu.* **1996**, *173*, 267.

27

Preparation of Dihydropyrimidinone using Green Chemistry Technique

Aim: To prepare dihydropyrimidinone using green chemistry technique.

Apparatus: Round bottomed flask, reflux condenser, beaker, Buchner assembly, etc.

Retrosynthetic Analysis:

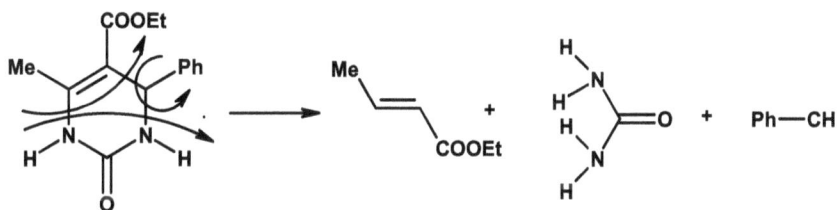

Dihydropyrimidine molecule can be split into urea, benzyl and ethyl ester of but-2-enoic acid as shown above. Urea can serve as one of the synthons. Benzaldehyde can act as a benzyl group donor and the tautomeric form of ethyl ester of butenoic acid is ethyl acetoacetate, thus can serve as synthons.

Chemicals:

Chemicals & Reagents	Mol. Wt.	Mol. For.	Moles	Quantity Taken
Ethylacetoacetate	114	$C_6H_{10}O_2$	0.011 M	1.3 g
Benzaldehyde	106	C_7H_6O	0.010 M	1.1 g
Urea	60	N_2H_4O	0.012 M	0.7 g

Reaction:

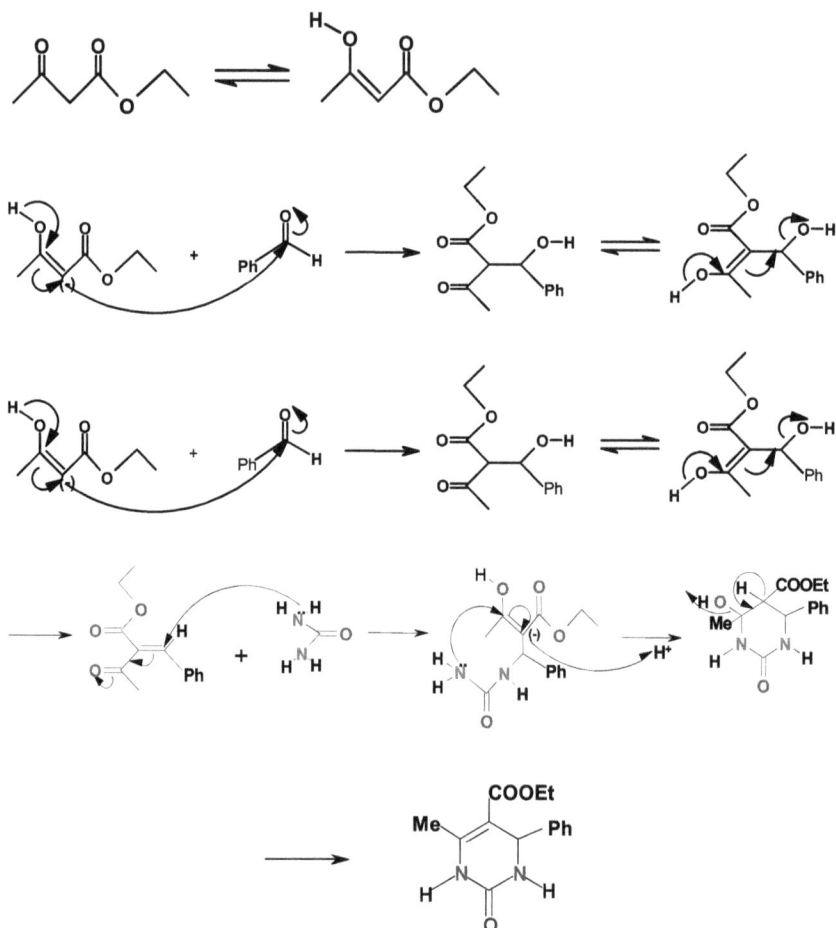

Principle:

Dihydropyrimidine is prepared by heating a mixture of ethylacetoacetate, benzaldehyde and urea in ethanol for 18 hours in the presence of a dehydrating agent, concentrated H_2SO_4. However this is a lengthy process involving the usage of sulphuric acid which is not eco friendly. Thus in this process the reaction is carried out under solvent-free condition without usage of sulphuric acid. The reaction completes within one hour in a rapid manner. Thus it is advantageous over conventional methods.

Procedure:

- Take benzaldehyde (1.1 g, 0.010 M), ethylacetoacetate (1.3 g, 0.011 M) and urea (0.7 g, 0.012 M) in a round bottomed flask and shake well for 2.0 minutes.

- Heat the reaction mixture at 90°C for 1.0 hours. With the progress of the reaction, solid starts depositing on the walls of the container and by the end of reaction the flask will be full of solid.

- Add crushed ice, filter the light yellow solid obtained under vacuum, wash with cold water, dry and recrystallize from rectified spirit to give a colorless product.

Physical Properties:

Percentage Yield: 75% $^W/_W$

Melting point: 201°C-202°C.

References:

1. Ranu, B. C.; Hajra, A.; Dey, S. S. *Org. Proc. Res. Dev.* **2002**, *6*, 817.

Preparations involving three step syntheses

28

Preparation of Sulfanilamide

(Synonyms: 4-Anilinesulfonamide, 4-Sulphamidoaniline)

Aim: To prepare sulfanilamide using acetanilide.

Apparatus: Round bottomed flask, water bath, reflux condenser, filter papers, funnel.

Retrosynthetic Analysis:

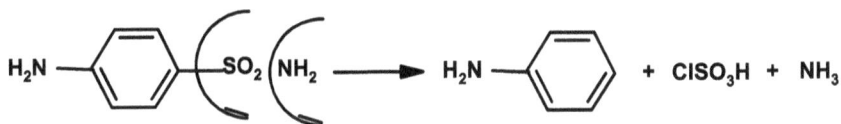

Ammonia and sulphurdioxide can be removed from sulphanilamide to give rise to aniline which can act as a substrate. Chlorosulphonic acid is a sulphurdioxide donor. Hence used as a reagent.

However it cannot react with aniline directly to give p-amino benzene sulphonyl chloride as amino (NH_2) group (due to its electron richness) reacts with chlorosulphonic acid before it reacts with benzene ring of aniline. When amino group is blocked by acetyl group it gives acetanilide. The acetanilide reacts with chlorosulphonic acid to give p–acetamido benzene sulphonyl chloride. Hence this finds its use as a synthon.

The chloro group in p–acetamido benzene sulphonyl chloride can be replaced by NH_2 by treating with ammonia as a reagent to give p-acetamido benzene sulphonamide. The hydrolysis of the same can result in sulphanilamide.

Chemicals:

Chemicals & Reagents	Mol. Wt.	Mol. Formula	Moles	Quantity Taken
Acetanilide	135	C$_8$H$_9$NO	0.019 M	2.5 g
Chloro sulphonic acid	116.5	ClSO$_3$H	0.05 M	6.25 g
Concentrated Ammonia	17	NH$_3$		3.75 ml
Dilute sulphuric acid	98	H$_2$SO$_4$		q.s.
Concentrated hydrochloric acid	36.5	HCl		5.0 ml
Sodiumbicarbonate	83.5	NAHCO$_3$		q.s.

Principle:

Sulphanilamide is also called as 4-amino benzene sulfonamide. The 4-acetamido benzene sulfonyl chloride is an essential intermediate in the synthesis of number of sulphonamides. It is prepared by chloro sulphonation of acetanilide. The formed p-acetamido benzene sulfonyl chloride readily reacts with ammonia and gives p-acetamido benzene sulfonamide. The acetamide group can be readily hydrolyzed and deacetylated in the presence of HCl to give sulphanilamide.

Reaction:

Step-I

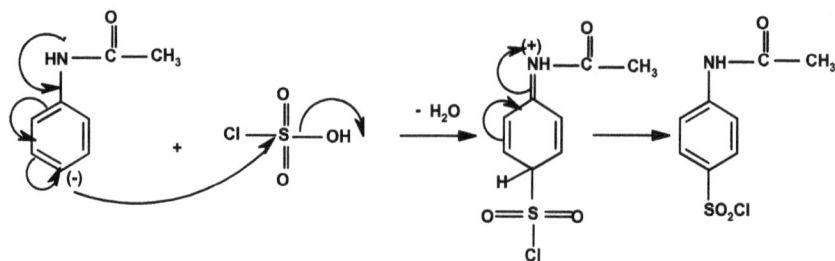

Step II

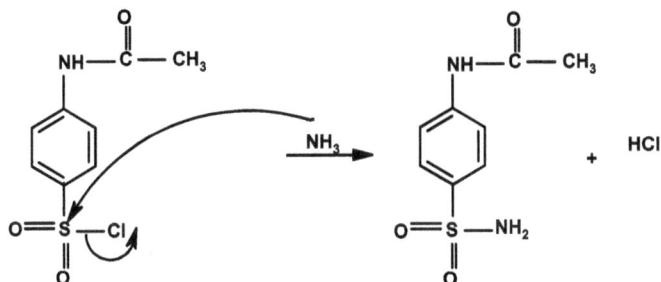

Step-III

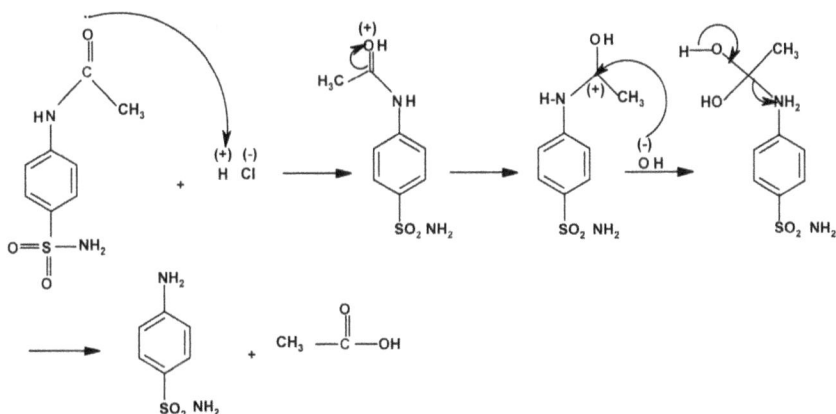

Procedure:

Preparation of 4-acetamido benzene sulfonyl chloride from acetanilide.

- Place a dry powder of acetanilide (2.5 g, 0.019 M) in a 100 ml round bottomed flask fitted with a reflux condenser and calcium guard tube.
- Add chlorosulphonic acid (6.25 g, 0.05 M) in small quantities with occasional shaking.
- Heat the reaction mixture under reflux for about 2.0 hours at 60°C – 70°C on a water bath.

- Cool the contents of the flask to room temperature and pour onto 100 g of crushed ice.

- Filter off the separated sulphonyl chloride derivative at the pump, wash thoroughly with water and dry.

Preparation of 4-acetamido benzene sulfonamide from p-acetamido benzene sulphonyl chloride:

- To 4-acetamido benzene sulphonyl chloride (product obtained from the first step) in a 100 ml beaker add concentrated ammonia (3.75 ml) with continuous stirring in a fuming cup board. A vigorous reaction takes place with evolution of heat.

- Stir the mixture until a smooth fine paste is obtained and then heat for about 30 minutes at $70^{\circ}C$ with occasional stirring.

- Cool the mixture to room temperature and neutralize with dilute H_2SO_4 until the contents are just acidic.

- Filter off the precipitated 4-acetamido benzene sulphonamide at the pump, wash with cold water and dry.

- The crude product obtained is sufficiently pure for using in the next step.

Preparation of sulphanilamide from 4-acetamido benzene sulphonamide.

- Place 4-acetamido benzene sulphonamide (product of second step) in a 100 ml round bottomed flask fitted with a reflux condenser, add 5.0 ml of concentrated hydrochloric acid and heat under reflux for 30 minutes.

- Decolorize the solution by charcoal treatment, cool to room temperature and filter.

- Add sodiumbicarbonate to the filtrate in small quantities with stirring until effervescence ceases.

- Filter the sulphanilamide precipitated at the pump, dry in an hot air oven and recrystallize from ethanol.

Physical Properties:

Melting Point: $166^{\circ}C$.

Percentage yield: 10% $^W/_W$

Category:

Antibacterial agent.

Identification test:

1. Sulphanilamide on heating with NaOH gives sulphanilic acid and evolves ammonia.

2. Can be diazotized and the diazonium salt couples with β-naphthol to form an azo dye.

References:

1. Gelmo, P. *J. Prakt. Chem.* **1908**, *77*, 369.

2. Galat, A. *Ind. Eng. Chem.* **1944**, *36*, 192.

3. Hurdis, Y. *J. Chem. Edu.* **1969**, *46*, 697.

29

Preparation of Trimethoprim

Aim: To prepare trimethoprim from 3,4,5-trimethoxy benzaldehyde and β-methoxypropionitrile.

Apparatus: Conical flask, water bath, round bottomed flask, reflux condenser, thermometer, Buchner assembly, electric water bath.

Retrosynthetic Analysis:

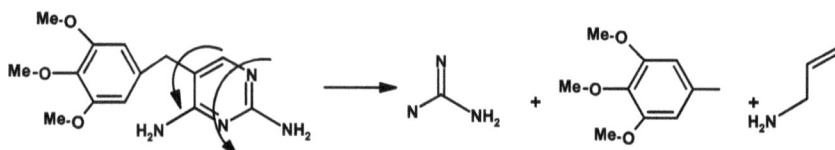

Trimethoprim molecule can be split as shown above. Guanidine can act as the synthon for fragment one. Trimethoxy benzaldehyde can be another synthon for fragment two and β-methoxy propionitrile can replace fragment three and thus can serve as another synthon.

Chemicals:

Chemical & Reagents	Mol. Wt.	Mol. For.	Moles	Quantity Taken
Sodium metal	23	Na	0.069	1.6 g
Absolute methanol	32	CH_4O		150 ml
β-Methoxy propionitrile	85	C_4H_7NO	0.055	4.75 g
3,4,5-Trimethoxy benzaldehyde	196	$C_{10}H_{12}O_4$	0.05	9.8 g
Benzene	78	C_6H_6		80 ml
Guanidine	59	CH_5N_3	0.0025	148 mg
Sulphuric acid (Aq. Soln.) (3N)	98	H_2SO_4		7 ml
Sodium hydroxide soln. (50 %)	40	NaOH		5 ml

Principle:

It involves 3 steps:

1. Preparation of 3, 4, 5-trimethoxy-2′-methoxymethyl-cinnamonitrile.
2. Preparation of 1′-(3,4,5-trimethoxy-phenyl)-2′-cyano-propyl methyl ether.
3. Preparation of trimethoprim.

Preparation of 3,4,5-trimethoxy-2′-methoxymethyl-cinnamonitrile from 3,4,5-trimethoxy benzaldehyde and β-methoxypropionitrile:

The β-methoxypropionitrile under strongly basic conditions of sodium methoxide gives rise to carbanion generation which reacts with 3,4,5-trimethoxy benzaldehyde to yield 3, 4, 5-trimethoxy-1-hydroxy 2′-methoxymethyl-propionitrile. It undergoes dehydration to give 3, 4, 5-trimethoxy-2′-methoxymethyl-cinnamonitrile.

Preparation of 1′- (3,4,5-trimethoxy-phenyl)-2′-cyano-propyl methyl ether from the above product:

The reduction of 3, 4, 5-trimethoxy-2′-methoxymethyl-cinnamonitrile gives rise to 1′- (3,4,5-trimethoxy-phenyl)-2′-cyano-propyl methyl ether. This is effected by sodium as reducing agent and methanol as solvent.

Preparation of trimethoprim from the above product:

Guanidine undergoes condensation with 1′- (3,4,5-trimethoxy-phenyl)-2′-cyano-propyl methyl ether under strongly basic conditions to give rise to trimethoprim.

Reaction:

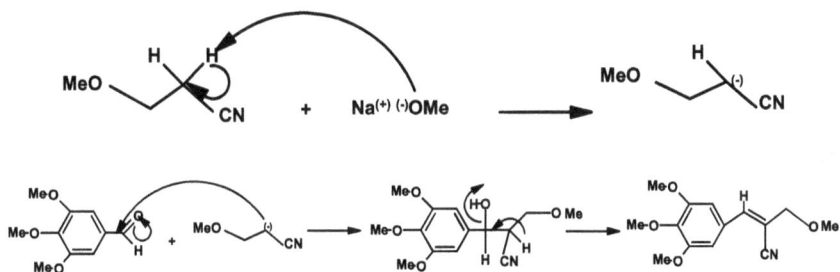

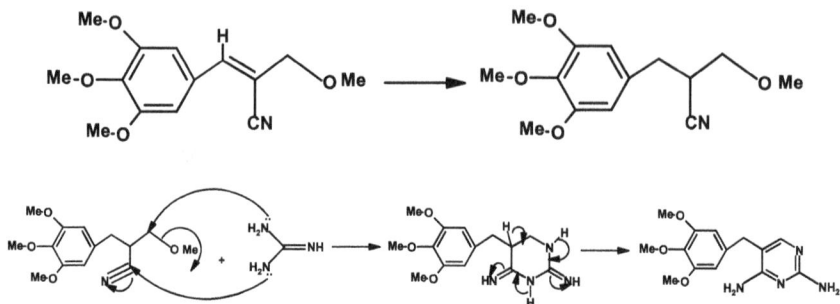

Procedure:

Step-1: Preparation of 3, 4, 5-trimethoxy-2′-methoxymethyl-cinnamonitrile.

- Cut sodium metal (0.6 g, 0.026 M), and dissolve in 30.0 ml of methanol under gentle stirring and refluxing.

- When most of the sodium metal dissolves, introduce carefully into the round bottom flask containing β-methoxypropionitrile (4.75 g, 0.055 M) and 3,4,5- trimethoxy benzaldehyde (9.8 g, 0.05 M) and heat the reaction mixture under reflux for 4 hours.

- Cool the resulting mixture to room temperature and then chill it in an ice-bath with the addition of 15.0 ml water into the reaction flask.

- The product crystallizes rapidly. Permit the crystallization to proceed at 5°C-10°C under gentle stirring for 60-70 minutes.

- Separate the crystallized product in a Buchner funnel under suction and wash with 20 ml of 60% ($^V/_V$) ice-cold methanol.

- Dry the crude product in air, so that it can be used for subsequent steps without any purification.

Physical Properties:

Melting Point: 78°C-80°C.

Percentage yield: 71% $^W/_W$

Step-2: Preparation of 1'-(3,4,5-trimethoxy-phenyl)-2´-cyano-propyl methyl ether.

- Cut sodium metal (1.9 g, 0.826 M) and dissolve in 30.0 ml of absolute methanol in a round bottom flask fitted with a reflux condenser and add 3,4,5-trimethoxy-2´-methoxymethyl cinnamon-nitrile (10.6 g, 0.04 M).

- Heat the reaction mixture under reflux for 24 hours. Carefully pour the solution, which had turned almost brown, into 100 ml of distilled water.

- Extract the precipitated oily residue successively with benzene.

- Wash the combined benzene layer (~ 60.0 ml) thoroughly 3 times with approximately 55.0 ml water.

- Remove benzene by distillation under reduced pressure on an electric water bath.

- Distill the obtained residual brown oil under vacuum. The clear, viscous oil 1'-(3,4,5-trimethoxy-phenyl)-2´-cyano-propyl methyl ether becomes solidifies upon standing.

- It can be obtained in pure form by recrystallization from methanol.

Physical Properties:

Melting Point: 69°C-70°C

Percentage yield: 70% $^W/_W$

Step-3: Preparation of Trimethoprim.

- Dissolve 3,4,5-trimethoxy-2´-cyano-dihydrocinnamaldehyde dimethyl acetal (3.15 g, 0.012 M), and guanidine (0.148 g, 0.032 M) in absolute methanol (20.0 ml) taken in a dry round bottom flask, and heat under reflux for 2 hours.

- Distill off the solvent completely under gentle stirring in an electric water bath.

- The yellowish crystalline mass solidifies almost completely.

- Cool the resulting residue and prepare slurry with 10.0 ml of distilled water.

- Filter it in a Buchner funnel under suction and dry subsequently under vacuum.

- Recrystallize the product by adding it to 3.0 ml aqueous H_2SO_4 (3.0 N) at 60°C under gentle stirring. Subsequently chill the solution to 5°C-10°C under stirring.

- Collect the crystalline sulphate salt by vacuum filtration and wash it twice with 1.0 ml of chilled 3.0 N aqueous H_2SO_4 each time.

- Dissolve the resulting sulphate product in 20.0 ml of hot distilled water, add activated charcoal powder and filter it.

- Gradually add the solution of 2.0 g of NaOH dissolved in 4.0 ml distilled water under thorough chilling until the desired product gets precipitated from the clear colorless filtrate.

- Filter the above solution by suction and wash it thoroughly with water on the filter paper until a white product is obtained.

Physical Properties:

Melting Point: 200°C-201°C.

Percentage yield: 90% $^W/_W$

Category:

- It finds its use as an antibacterial agent against a wide spectrum of organisms.

- It is used widely in combination with sulphamethoxazole (Cotrimoxazole).

- The combination of dapsone and trimethoprim is used in the treatment of leprosy and infections caused by *Mycobacterium avium*.

References:

1. Stenbuck, H. *US 3049544*.

2. Hoffer, *US 3341541*.

3. Roth, B. et al., *J. Med. Chem.* **1980**, *379*, 535.

Part D

Quantitative Estimation of Drugs/API's in Formulations

Assay of Sulphaguanidine Tablets

Aim: To carry out the assay of sulphaguanidine tablets and report the percentage label claim.

Apparatus: Conical flask, burette, pipette, stand, tile, glass rod.

Chemicals:

Chemicals & Reagents	Quantity Taken
Sulphaguanidine tablets	20
Sodium nitrite solution (0.1 M)	*q.s.*
Hydrochloric acid solution (2.0 M)	15.0 ml
Starch iodide suspension	Few ml

Principle:

The principle involved in assay of sulphaguanidine tablets is diazotization titrimetrically. Aromatic primary amines like sulpha guanidine containing free amino groups react with nitrous acid to form diazonium salts at 0°C-5°C. Nitrous acid is prepared *in situ* by reacting sodium nitrite with concentrated hydrochloric acid. The observation of end point depends upon small excess of sodium nitrite which is then present and can be visually demonstrated using starch iodide paper as an external indicator. The liberated iodine reacts with starch to produce blue color.

Reaction:

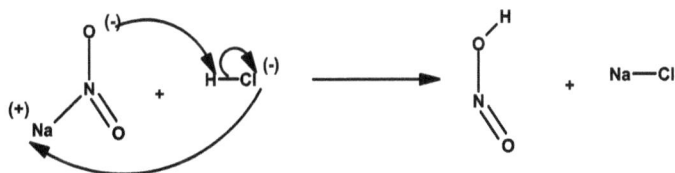

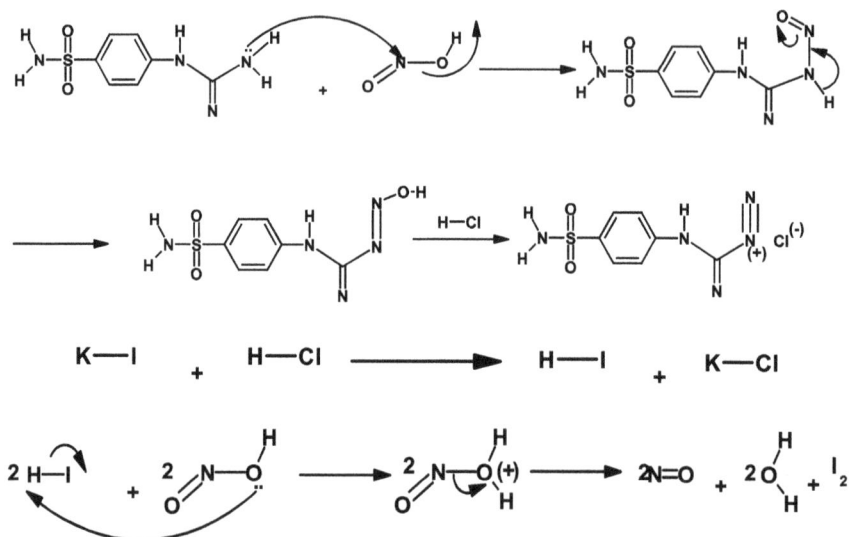

Preparation of 0.1 M sodium nitrite solution:

Weigh accurately about 6.9 g of solid sodium nitrite and dissolve in few ml of water.

Dilute the resulting solution up to 100 ml with distilled water.

Procedure for standardization of 0.1 M sodium nitrite:

- Accurately weigh about 0.1 g of sulphanilic acid into a conical flask, add 1.5 g of potassium bromide followed by 15.0 ml of 2.0 M hydrochloric acid (HCl) and cool to below 10°C.

- Titrate the mixture with 0.1 M sodium nitrite solution at a temperature not exceeding 10°C until a drop of solution immediately gives blue color when drawn quickly by means of a fine glass rod across the surface of starch paper.

- Titration is complete when the end point is reproducible after the titrated solution is allowed to stand for 1 minute.

- Each ml of 0.1 M sodium nitrite ($NaNO_2$) is equivalent to 0.01732 g of sulphanilic acid.

Procedure for assay:

- Weigh a set of 20 tablets accurately and reduce it to a fine powder.

- Weigh accurately a quantity of powder equivalent to 0.05 g of sulphaguanidine into a dry and clean conical flask.

- Add a mixture of water (15.0 ml) and concentrated HCl (2.0 ml) to the above tablet powder and cool to below 10°C.

- Titrate the mixture with 0.1 M sodium nitrite solution maintaining the temperature NMT 15°C until a drop of solution immediately gives blue color when drawn across surface of starch iodide paper.

- Repeat the titration several times until a reproducible end point is achieved.

- Equivalent Factor: Each ml of 0.1 M sodium nitrite is equivalent to 0.02323 g of sulphaguanidine.

Category:

Antibacterial, para amino benzoic aicd (PABA) antagonist.

Standards:

Sulphaguanidine tablet contain NLT 95% and NMT 105% of standard amount of sulphaguanidine.

References:

1. Beckett, A. H.; Stenlake, J. B. *Practical Pharmaceutical Chemistry*, CBS Publishers and Distributors, New Delhi, 4th Ed., Vol 1, pp 257.

## 2

Assay of Dapsone Tablets

Aim: To estimate the amount of dapsone present in the given sample tablets.

Apparatus: Burette, pipette, sand, glass rod, ice bath, tile.

Chemicals:

Chemicals & Reagents	Quantity Taken
Dapsone tablets	20
Sodium nitrite solution (0.1 M)	*q.s.*
Hydrochloric acid solution (2.0 M)	6.0 ml
Starch Iodide suspension	Few milli litres

Principle:

The principle involved in the assay of dapsone tablets is diazotization. Aromatic primary amines like dapsone react with nitrous acid to form diazonium salts. Nitrous acid is prepared *in situ* by reacting sodium nitrite with concentrated hydrochloric acid at 0°C-5°C. The observation of the end point depends on detection of small excess of nitrous acid present at that point which can be visually determined by using starch iodide suspension as an external indicator. The iodine liberated reacts with starch to produce blue color.

Reaction:

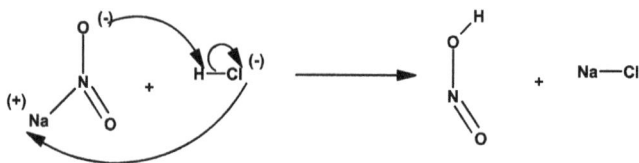

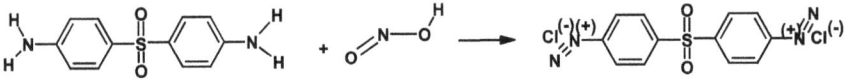

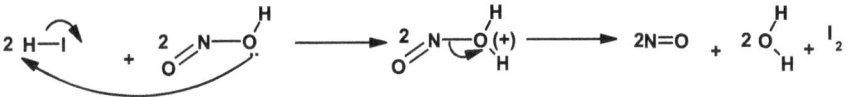

Preparation of 0.1 M sodium nitrite solution:

- Weigh accurately about 6.9 g of sodium nitrite and dissolve in few ml of water.
- Dilute the resulting solution up to 100 ml with distilled water.

Procedure for standardization of 0.1 M sodium nitrite:

- Accurately weigh about 0.1 g of sulphanilic acid into a conical flask, add 1.5 g of potassium bromide followed by 15.0 ml of 2.0 M HCl and cool to below 10°C.
- Titrate the mixture with 0.1 M sodium nitrite solution at a temperature not exceeding 10°C until a drop of solution immediately gives blue color when drawn quickly by means of a fine glass rod across the surface of starch paper.
- Titration is complete when the end point is reproducible after the titrated solution is allowed to stand for 1 minute.
- Each ml of 0.1 M $NaNO_2$ is equivalent to 0.01732 g of sulphanilic acid.

Procedure for assay:

- Weigh 10 tablets of dapsone accurately and reduce it to a fine powder with the help of a mortar and pestle.
- Weigh a quantity of powder equivalent to 0.1g of dapsone into a conical flask and dissolve in a mixture of 6.0 ml of water and 6.0 ml of 2 M HCl.
- Cool the mixture to below 10°C and titrate against 0.1 M sodium nitrite at a temperature not exceeding 10°C until a drop of solution gives dark blue color when drawn quickly across the surface of starch iodide paper with a glass rod.

- The titration is complete when the end point is reproducible after allowing the solution to stand for 1 minute.

- Equivalent factor: Each ml of 0.1 M sodium nitrite is equivalent to 0.01242g of Dapsone.

Standards:

Dapsone tablets contain NLT 93 % and NMT 107 % of stated amount of dapsone.

References:

1. Indian Pharmacopoeia 1996 Vol I, Ministry of Health and Family Welfare, Govt. of India pp 223.

2. Indian Pharmacopoeia 1996 Vol II, Ministry of Health and Family Welfare, Govt. of India Appendix 213.

3. Beckett, A. H.; Stenlake, J. B. *Practical Pharmaceutical Chemistry*, CBS Publishers and Distributors, New Delhi, 4th Ed., Vol 1, pp 258.

## 3

Assay of Chloroquine Phosphate

Aim: To estimate the amount of chloroquine phosphate present in the given sample.

Apparatus: Burette, pipette, tile, conical flask, etc.

Chemicals:

Chemicals & Reagents	Quantity Taken
Chloroquine phosphate	0.05 g
Glacial acetic acid	50 ml
Potassium hydrogen phthalate	0.3 g
Crystal violet	Few drops
Perchloric acid	*q.s.*

Principle:

The assay of chloroquine phosphate is carried out by nonaqueous titrimetry. Chloroquine phosphate is chemically RS-4-(7-chloro-quinolinylamine)pentyldiethylamine diphosphate. It exists in 2 isomeric forms. It is a weak base having a pH of 4.8-6.0. They do not give sharp end point when titrated in aqueous medium. So it is analyzed by nonaqueous method.

In nonaqueous titration, nonaqueous solvents such as acetic acid (CH_3COOH) (100%), acetone and weak organic bases, like tetrabutylammonium hydroxide in methanol, potassium, sodium, or lithium methoxide in toluene-methanol, are used for titration of weekly basic and weekly acidic substances respectively.

Weekly acidic and basic substances do not give sharp end points in aqueous titrimetry due to its low dissociation constant. The amphiprotic solvent, like acetic acid, which is used as a solvent in nonaqueous titrations of basic substances functions both as an acid and as a base.

When it is mixed with perchloric acid, it acts as a base and accepts proton. In the presence of a base like pyridine, it acts as an acid by donating protons. Thus it increases the acidity or basicity of the weekly acidic or basic substances hence enhances the attainment of sharp endpoint.

In the preparation of perchloric acid (0.1 M) solution, acetic anhydride is added to absorb the moisture in the perchloric acid and acetic acid, rendering it virtually anhydrous. Though excess of acetic anhydride is not always disadvantageous, care must be taken to avoid excess when primary and secondary amines have to be titrated. It readily acylate them to give neutral and reduce its basicity products. The perchloric acid must be well diluted with acetic acid before adding the acetic anhydride. If it is not done it may lead to explosive acetyl perchlorate.

Potassium hydrogen phthalate is used as a primary standard for standardization of perchloric acid.

Reaction:

$$HClO_4 + CH_3COOH \longrightarrow CH_3COOH_2^{(+)} + ClO_4^{(-)}$$

$$CH_3COO^{(-)} + CH_3COOH_2^{(+)} \longrightarrow CH_3COOH + CH_3COOH$$

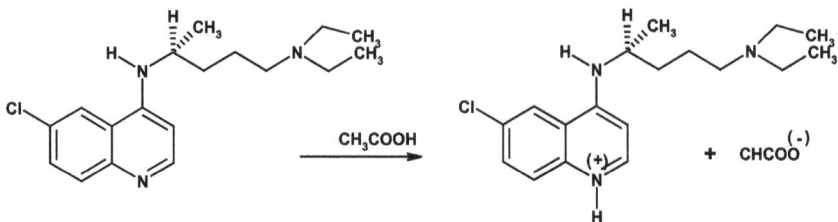

Overall Reaction:

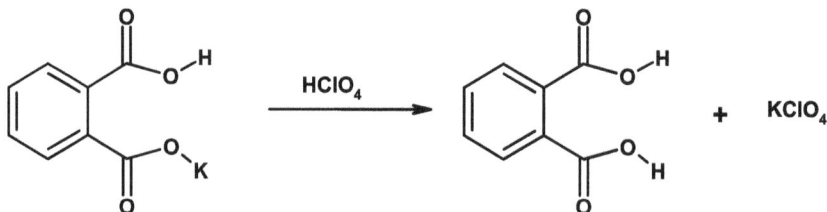

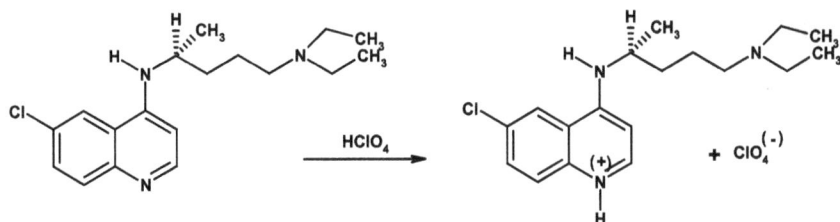

Preparation of 0.1 M perchloric acid:

- Mix an accurately measured volume of Perchloric acid (8.5 ml) with anhydrous glacial acetic acid (500.0 ml).
- Then add 25.0 ml of acetic anhydride.
- Cool the mixture to room temperature and make up the volume to 1000 ml using anhydrous glacial acetic acid.
- Allow the solution to stand for over night.

Standardization of 0.1 M perchloric acid:

- Accurately weigh 0.1 g of potassium hydrogen phthalate and dissolve it in 10.0 ml of glacial acetic acid.
- Add a few drops of crystal violet solution and titrate with 0.1 M perchloric acid until the color changes from violet to emerald green.
- Perform a blank determination and make necessary corrections.

Procedure for assay:

- Weigh accurately around 0.05 g of chloroquine phosphate into a conical flask and dissolve in 10.0 ml of anhydrous glacial acetic acid with the aid of heat if necessary.
- Add 0.1 ml of crystal violet and titrate the resulting solution with 0.1 M perchloric acid until the color changes from violet to emerald green.
- Perform a blank titration and incorporate necessary corrections.
- Equivalent factor: Each ml of 0.1 M perchloric acid is equivalent to 0.02579 g of chloroquine phosphate.

Category:

Antimalarial, antiamoebic.

Reference:

1. Indian Pharmacopoeia, Vol-1 Ministry of Health and Family Welfare, Govt. of India 1996 pp. 172.

2. Indian Pharmacopoeia, Vol-II Ministry of Health and Family Welfare, Govt. of India Appendix 3.45

3. Beckett, A. H.; Stenlake, J. B. *Practical Pharmaceutical Chemistry*, CBS Publishers and Distributors, New Delhi, 4th Ed., Vol 1, pp 233-234.

4

Assay of Metronidazole Tablets

Aim: To estimate amount of metronidazole present in the given sample.

Apparatus: Burette, pipette, tile, conical flask etc.

Chemicals:

Chemicals & Reagents	Quantity Taken
Metronidazole tablets	20
Glacial acetic acid	50 ml
Potassium hydrogen phthalate	0.3 g
Crystal violet	Few drops
Perchloric acid	*q.s.*

Principle:

Chemically metronidazole is 2-[3-methyl(5-nitro-1H-imidazole)-1-yl] ethanol. The principle involved in the assay of metronidazole tablets is nonaqueous titration using perchloric acid. Metronidazole is a weak base and gives poor end point in aqueous titration. However the end point is satisfactory in nonaqueous titration in the presence of solvents like glacial acetic acid. it is an amine and accepts a proton from perchloric acid to give the corresponding conjugate acid based on Brownstead Lowry theory of acids and bases. Crystal violet is used as an indicator to detect the end point.

In nonaqueous titration, non aqueous solvents such as CH_3COOH (100%), acetone and weak organic bases, like tetrabutylammonium hydroxide in methanol, potassium, sodium, or lithium methoxide in toluene-methanol, are used for titration of weakly basic and weakly acidic substances respectively.

Weakly acidic and basic substances do not give sharp end points in aqueous titrimetry due to its low dissociation constant. The amphiprotic

231

solvent, like acetic acid, which is used as a solvent in non aqueous titrations of basic substances functions both as an acid and as a base. When it is mixed with perchloric acid, it acts as a base and accepts proton. In the presence of a base like pyridine, it acts as an acid by donating protons. Thus it increases the acidity or basicity of the weekly acidic or basic substances hence enhances the attainment of sharp endpoint.

In the preparation of perchloric acid (0.1 M) solution, acetic anhydride is added to absorb the moisture in the perchloric acid and acetic acid rendering it virtually anhydrous. Though excess of acetic anhydride is not always disadvantageous, care must be taken to avoid excess when primary and secondary amines have to be titrated. It readily acylates them, to give neutral products and reduce its basicity. The perchloric acid must be well diluted with acetic acid before adding the acetic anhydride. If it is not done it may lead to explosive acetyl perchlorate.

Potassium hydrogen phthalate is used as a primary standard for standardization of perchloric acid.

Reaction:

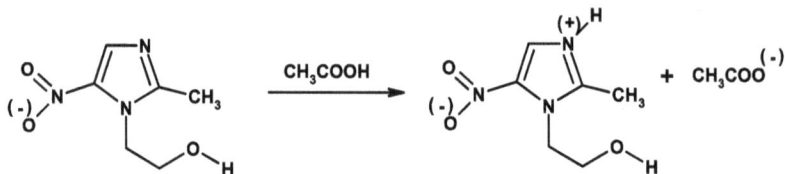

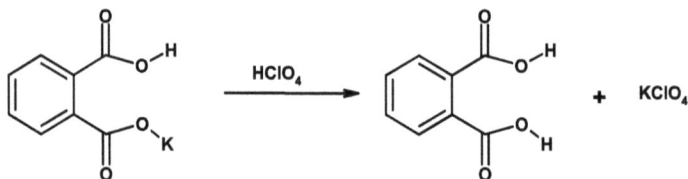

Overall Reaction:

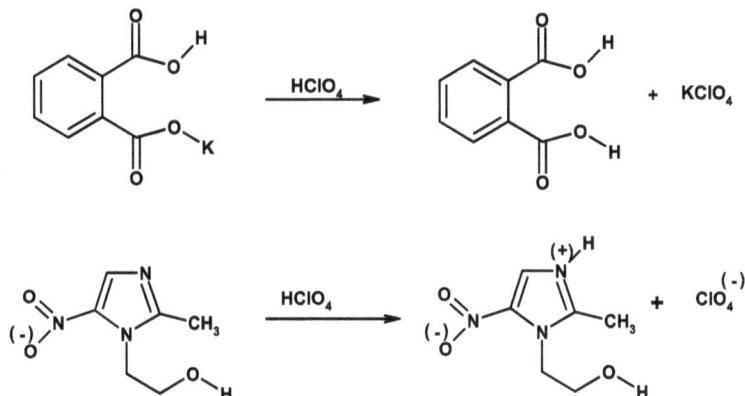

Preparation of 0.1 M perchloric acid:

- Mix an accurately measured volume of perchloric acid (8.5 ml) with anhydrous glacial acetic acid (500.0 ml).
- Then add 25.0 ml of acetic anhydride.
- Cool the mixture to room temperature and make up the volume to 1000 ml using anhydrous glacial acetic acid.
- Allow the solution to stand for over night.

Standardisation of 0.1 M perchloric acid:

- Accurately weigh 0.1 g of potassium hydrogen phthalate and dissolve it in 10.0 ml of glacial acetic acid.
- Add a few drops of crystal violet solution and titrate with 0.1 M perchloric acid until the color changes from violet to emerald green.
- Perform a blank determination and make necessary corrections.
- Equivalent factor: Each ml of 0.1 M perchloric acid = 0.02042 g of potassium hydrogen phthalate.

Procedure:

- Accurately weigh 20 tablets of metronidazole and reduce them to a fine powder.
- Weight out a quantity of powder equivalent to 0.1 g of metronidazole into a dry conical flask.

- Add glacial acetic acid (15.0 ml) followed by a drop of crystal violet solution to the above conical flask and dissolve.
- Titrate the resulting solution against perchloric acid until the color changes from violet to emerald green.
- Perform a blank titration and incorporate necessary corrections.
- Equivalent factor: Each ml of 0.1 M perchloric acid ($HClO_4$) is equivalent to 0.01712 g of metronidazole.

Category:

Anti amoebic, antibacterial, antiprotozoal.

Standards: Each tablet contains NLT 95% and NMT 105% of the stated amount of metronidazole.

Reference:

1. Indian Pharmacopoeia, Vol-1 1996 Ministry of Health and Family Welfare, Govt. of India pp. 489.
2. Indian Pharmacopoeia, Vol-II Ministry of Health and Family Welfare, Govt. of India Appendix 3.45.
3. Beckett, A. H.; Stenlake, J. B. *Practical Pharmaceutical Chemistry*, CBS Publishers and Distributors, New Delhi, 4th Ed., Vol 1, pp 167-168.

5

Assay of Diethyl Carbamazine Citrate Tablets

Aim: To estimate the amount of diethyl carbamazine present in the given sample.

Apparatus: Conical flask, burette, pipette, stand, tile etc.

Chemicals:

Chemicals & Reagents	Quantity Taken
Diethyl carbamazine citrate tablets	20
Potassium hydrogen phthalate	0.5 g
Sodium hydroxide solution (5.0 M)	10.0 ml
Sulphuric acid (0.05 M)	15.0 ml
Chloroform	20.0 ml
Sodium hydroxide (0.1 M)	*q.s.*
Bromocresol green	Few drops

Principle:

Chemically diethyl carbamazine [DEC] is 1,1-diethyl-carbamoyl-4-methyl-piperazine or N, N-diethyl-4-methyl piperazine. it is official in I.P in the form of its freely soluble citrate. It is estimated by means of 0.05 M sulphuric acid and back titration with 0.1 M sodium hydroxide using bromocresol green as an indicator. Preliminary hydrolysis of tablet powder converts diethyl carbamazine citrate to diethyl carbamazine and sodium citrate. The free diethyl carbamazine is extracted into chloroform and the base is converted into its sulphate salt by treating with H_2SO_4. Finally the diethyl carbamazine is dissolved in water and the aqueous layer is back titrated with standard base using bromocresol green as indicator until the color changes from yellow to bluish green.

Reaction:

$$H_2SO_4 \ + \ 2\,NaOH \longrightarrow Na_2SO_4 \ + \ 2H_2O$$

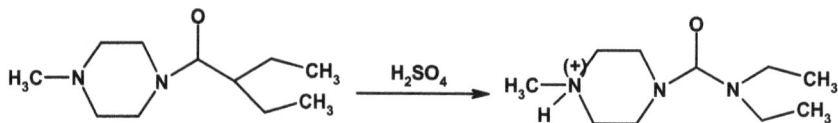

Preparation of 0.1 M NaOH

- Weigh accurately sodium hydroxide (4.0 g, 0.1 M) and dissolve in few millilitres of water and make up the volume to 1000 ml with distilled water.

Procedure for standardization of 0.1 M NaOH

- Weigh out accurately potassium hydrogen phthalate [PHP] (0.1 g) into a clean and dried conical flask and dissolve in water (10.0 ml).
- Titrate the resulting solution with 0.1 M NaOH using phenolphthalein as indicator until the appearance of pale pink color.
- Each ml of 0.1 M NaOH=0.020422 g of PHP.

Procedure for assay:

- Accurately weigh 20 tablets of DEC and reduce to a fine powder with the help of mortar and pestle.
- Weigh a quantity of powder equivalent to 0.1 g of DEC into a clean, dry conical flask and dissolve in 10.0 ml of 5.0 M NaOH as completely as possible.
- Extract the solution with 20.0 ml of chloroform and wash the chloroform extract continuously with water until the aqueous layer is not alkaline to phenolphthalein.
- Then extract the chloroform layer in succession with 15.0 ml of each of 0.05 M H_2SO_4. Combine the aqueous extracts.

- Titrate the resulting solution with 0.1M NaOH using bromocresol green as indicator until the appearance of blue colour.

- Equivalent factor: Each ml of 0.05M H_2SO_4 is equivalent to 0.03914 g of diethyl Carbamazine Citrate.

Category: Anthelmintic, antifilarial.

Limits: DEC tablets contains NLT 92.5% and NMT 107.5% of the stated amount of DEC.

Reference:

1. Indian Pharmacopoeia   1996 volume-1 Ministry of Health and Family Welfare, Govt. of India pp 248.

## 6

Assay of Isoniazid

Aim: To estimate the amount of isoniazid present in the given sample.

Apparatus: Conical flask, burette, pipette, stand, tile, etc.

Chemicals:

Chemicals & Reagents	Quantity Taken
Isoniazid tablets	20
Potassium bromate	*q.s.*
Potassium bromide	0.2 g
Methyl red	Few drops

Principle:

Chemically isoniazid is isonicotinic acid hydrazide. The principle involved in the assay of isoniazid tablets is redox titration. it is oxidized to isonicotinic acid by bromine in the presence of HCl to give hydrogen bromide with the evolution of nitrogen. Bromine is released as titration proceeds and reacts with isoniazid but the solution of bromine is not stable. Therefore to the solution potassium bromide is added which is slowly titrated with potassium bromate. The reaction between potassium bromide, HCl and potassium bromate releases bromine which oxidizes isoniazid. At the end point the HCl is depleted, where the methyl red solution changes from red to yellow.

Reaction:

$$KBrO_3 + 5\,KBr + 6\,HCl \longrightarrow 3\,Br_2 + 6\,KCl + 3\,H_2O$$

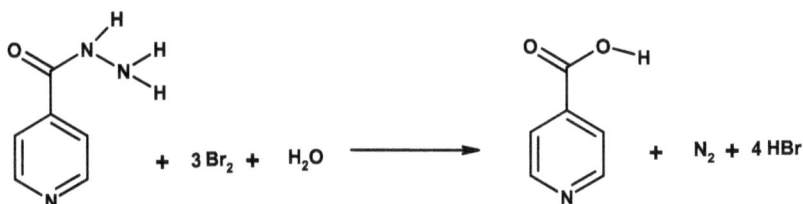

Preparation of 0.0167 M potassium bromate:

- Weigh accurately potassium bromate (2.783 g, 0.0167 M) and dissolve in minimum quantity of distilled water.
- Make up the volume up to 100 ml using distilled water.

Procedure for assay:

- Accurately weigh 20 tablets and reduce it to fine powder in a mortar and pestle.
- Weigh a quantity of tablet powder equivalent to 0.4 g of isoniazid and dissolve in 50.0 ml of water as completely as possible.
- Filter the solution and wash the residue several times and mix with the original solution. Make up the filtered solution to 250.0 ml with water.
- To 50.0 ml of the above solution, add 50.0 ml of water and 0.2 g of potassium bromide and 10.0 ml of HCl.
- Titrate the resulting solution slowly with 0.0167 M potassium bromate with continuous stirring using methyl red solution as indicator until the color changes from red to yellow.
- Equivalent factor: Each ml of 0.0167 potassium bromate is equivalent to 0.00342 g of isoniazid.

Category:

Antitubercular agent.

Standards: Isoniazid tablet contains NLT 90% and NMT 105% of the stated amount of isoniazid.

Reference:

1. Indian Pharmacopoeia, 1996, Vol-I Ministry of Health and Family Welfare, Govt. of India. pp 408.

2. Beckett, A.H.; Stenlake, J.B. *Practical Pharmaceutical Chemistry*, CBS Publishers and Distributors, New Delhi, 4th Ed., Vol 1, pp 187, 193-194.

7

Assay of Calcium Gluconate Injection

Aim: To estimate the amount of calcium gluconate present in the given sample.

Apparatus: Beaker, pipette, burette, conical flask, burner, etc.

Chemicals:

Chemicals & Reagents	Quantity Taken
Calcium gluconate	0.2 g
Ammonia buffer	5.0 ml
Mordant black-II	2 drops
Magnesium sulphate solution (0.05 M)	5.0 ml
Disodium ethylene diaminetetraacetate (0.05 M)	*q.s.*

Principle:

Chemically calcium d-gluconate monohydrate. Calcium gluconate is estimated by compleximetric titration using disodium edetate as complexing agent. Calcium ion of calcium gluconate forms a stable complex with disodium edetate and gives a water soluble metal complex. The formed calcium complex is stabilized by adjusting the pH with strong ammonium solution. Metal ions like Ca^{2+} give poor end point if titrated directly. In such cases other metals like magnesium in the form of magnesium sulphate is added to increase the sensitivity of end point. A metal ion indicator like mordant black-II mixture is used to detect the end point. It forms loose complexes with calcium ions and shows wine-red color during titration and turns blue at the end. A known volume of magnesium sulphate is added to the sample before titration for improvised visualization of the end point. Mg^{2+} ions are not complexed with disodium edetate in presence of calcium ion since calcium EDTA complex is more stable. When all the calcium ions have been complexed, the magnesium indicator complex breaks liberating magnesium which forms a complex with disodium edetate. From the titrated value the volume of magnesium sulphate ($MgSO_4$) added is subtracted.

Reaction:

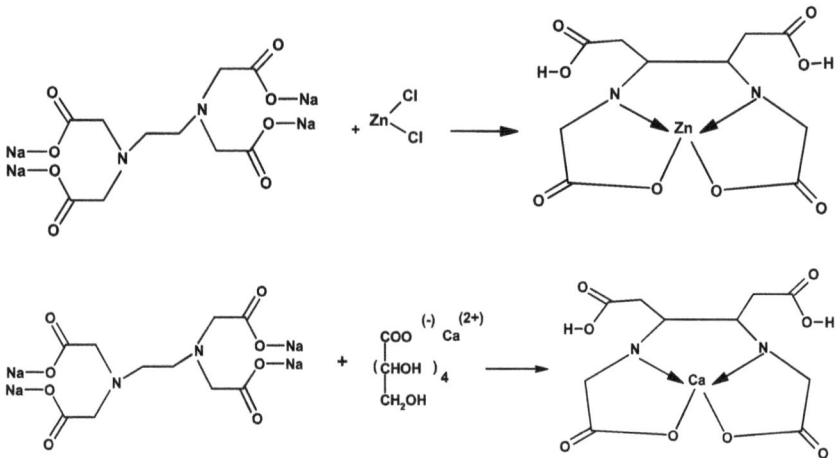

Preparation of 0.05 M disodium edetate:

- Dissolve 18.6 g of disodium edetate in 20.0 ml of water.
- Make up the volume to 100.0 ml using distilled water.

Preparation of mordent black-II mixture:

- Mix mordant black-II and NaCl in the ratio of 1:99 respectively.

Procedure for standardization of EDTA:

- Accurately weigh granulated zinc (0.03 g, 0.05 M) and dissolve in 5.0 ml of concentrated HCl.

- Add bromine water (0.1 ml) and warm the mixture slightly to remove excess bromine. Neutralize the resultant solution with 2 M NaOH and add sufficient amount of ammonia to adjust to pH-10 to dissolve the precipitate.

- To this solution add 50 mg of mordant black-II indicator and titrate against disodium edetate until color changes from wine-red to blue.

- Equivalent factor: Each ml of 0.05 M EDTA=0.0325g of zinc.

Procedure for Assay:

- Accurately weigh calcium gluconate (0.2 g) and dissolve in 25.0 ml of water. Add 0.05 M magnesium sulphate (5.0 ml) and strong ammonia solution (5.0 ml).

- Titrate the resulting solution against 0.05 M disodium edetate using mordant black-II mixture as indicator until color changes from wine-red to blue.

- From the volume of EDTA consumed, subtract the volume of magnesium sulphate added and calculate to get the results.

- Equivalent factor: Each ml of the remainder of 0.05 M EDTA is equivalent to 0.02242 g of calcium gluconate.

Category:

- Calcium replenisher.

- Calcium controls the excitability of nerves and regulates permeability of cell membrane and also maintains integrity of cell membrane.

- Calcium is essential for exciting and contraction of skeletal muscle. Excitation of exocrine and endocrine glands.

- Calcium is extracellular messenger for hormones, autocoids, neurotransmitters.

- Calcium is responsible for impulse generation in heart and also involved in coagulation of blood.

Standards: The given sample containing NLT 97.2% and NMT 102% of standard amount of calcium gluconate.

Reference

1. Indian Pharmacopoeia 1996 Vol-1 Ministry of Health and Family Welfare, Govt. of India pp 128.

2. Beckett, A. H.; Stenlake, J. B. *Practical Pharmaceutical Chemistry*, CBS Publishers and Distributors, New Delhi, 4th Ed., Vol 1, pp 219-220.

## 8

Assay of Furosimide Tablets

Aim: To estimate the amount of furosimide in the given sample.

Apparatus: Conical flask, burette, stand, pipette, tile, etc.

Chemicals:

Chemicals & Reagents	Quantity Taken
Furosimide tablets	20
Dimethylformamide	5.0 ml
Bromothymol blue	Few drops
Sodium hydroxide	*q.s.*

Principle:

Chemically furosimide is 4-chloro-N-furfuryl-5-sulphonyl anthranilic acid. The principle involved in the assay of furosimide is simple acid base titration. Here furosimide is weak acid which is titrated with a strong base like NaOH in the presence of a protophylic solvent like dimethylformamide (DMF). DMF enhances the acidity of furosimide and thus it can be titrated with strong base. Bromothymol blue is used as an indicator to detect the end point. To nullify the effect of solvent a blank determination was needed to make necessary corrections.

Reaction:

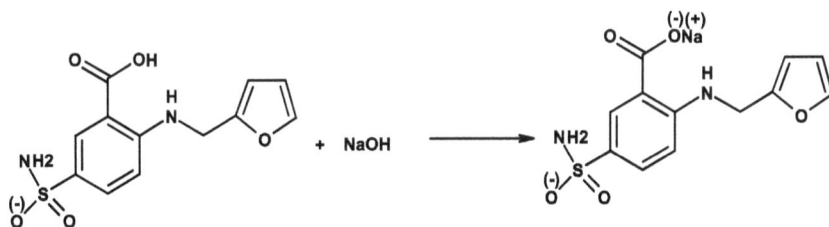

245

Preparation of 0.1 M NaOH

- Weigh accurately sodium hydroxide (4.0 g, 0.1 M) and dissolve in few millilitres of water and make up the volume to 1000 ml with distilled water.

Procedure for standardization of 0.1 M NaOH

- Weigh out accurately potassium hydrogen phthalate [PHP] (0.1 g) into a clean and dried conical flask and dissolve in water (10.0 ml).
- Titrate the resulting solution with 0.1 M NaOH using phenolphthalein as indicator until the appearance of pale pink color.
- Each ml of 0.1 M NaOH=0.020422g of PHP.

Procedure for assay:

- Accurately weigh 10 tablets of furosimide and reduce it to fine powder.
- Weigh a quantity of tablet powder equivalent 0.05 g of furosimide accurately into a clean conical flask and dissolve in 5.0 ml of dimethylformamide.
- Add a drop of bromothymol blue and titrate against 0.1M NaOH until the color changes from yellow to blue.
- Perform a blank titration and incorporate necessary corrections.
- Equivalent factor: Each ml of 0.1 M NaOH is equivalent to 0.03308 g of furosimide.

Category: Diuretic.

Standards: Each tablet contains NLT 90% and NMT 10% of the stated amount of furosimide.

Reference:

1. Indian Pharmacopoeia, 1996, Vol-I, Ministry of Health and Family Welfare, Govt. of India pp 333.

9

Assay of Aminophyline Injection

Aim: To conduct an assay for the given sample of aminophyline injection.

Apparatus: Conical flask, burette, pipette, stand etc.

Chemicals:

Chemicals & Reagents	Quantity Taken
Aminophyline	0.5 g
Silver nitrate (0.1 M)	20.0 ml
Nitric acid	*q.s.*
Ferric ammonium sulphate	2.0 ml
Ammonium cyanate	*q.s.*
Hydrochloric acid (0.1 M)	*q.s.*
Methyl red	Few drops

Principle:

Aminophylline is a stable mixture or combination of 2 molecules of the theophylline and 1 molecule of ethylenediamine. In the assay both theophylline and ethylenediamine are estimated separately. Theophylline content is determined by argentimetric titration. Theophylline is a weakly acidic substance therefore rapidly does not lose its proton. Hence silver nitrate is used in which nitrate extracts the proton and acts as a bronsted acid. It converts theophylline into silver salt with titration of equivalent amount of nitric acid. This acid is neutralized with a stronger base and excess of silver nitrate is estimated by titrating with ammonium thiocyanate solution. Ethylenediamine is determined by titration with acid

0.1 M HCl by using methyl orange as an indicator. The titration of weak base with strong mineral acid give its salts which will dissolve in aqueous solution to greater or lesser extent depending on dissociation constant of base. The pH of solution at the equivalent point will be less than 7 and therefore methyl orange is used as indicator.

Reaction:

$$AgNO_3 + NH_4SCN \longrightarrow AgSCN + NH_4NO_3$$

Preparation of ferric ammonium sulphate:

- Dissolve an accurately weighed ferric ammonium sulphate (8.0 g) in sufficient water to produce 100 ml.

Preparation of eosine red solution:

- Dissolve eosin red solution (0.5 g) in 10.0 ml of water.
- Then make up the volume to 100.0 ml with water.

Preparation of 0.1 M silver nitrate:

- Dissolve silver nitrate (17.0 g) in sufficient quantity of water to produce 100.0 ml.

Preparation of 0.1 M ammonium thiocyanate (NH₄SCN):

- Dissolve ammonium thiocyanate (7.612 g) in sufficient water to produce 100.0 ml.

Preparation of methyl red indicator:

- Dissolve methyl red (0.05 g) in a mixture of 0.1M NaOH (1.86 ml) and 95% ethanol (50 ml).
- Then make up the volume to 100 ml using distilled water.

Procedure for standardization of 0.1 M silver nitrate (AgNO₃) solution:

- Weigh accurately sodium chloride (0.1 g) (previously dried at 110°C for 2 hours) and dissolve in 5.0 ml of water in a conical flask.
- To this add acetic acid (5.0 ml), methanol (10.0 ml) and eosin solution (0.15 ml) and stir.
- Titrate the resulting mixture with 0.1 M AgNO₃ until color changes from pink to reddish violet.

Standardization of 0.1 M NH₄SCN:

- Pipette out 0.1 M AgNO₃ (30.0 ml) into a glass stopperd flask and dilute to 50.0 ml with distilled water.
- To the above solution, add nitric acid (2.0 ml) and ferric ammonium sulphate (2.0 ml).
- Titrate the resulting solution with 0.1 M ammonium thiocyanate solution until the first appearance of reddish brown color.
- Equivalent factor: Each ml of 0.1 M AgNO₃=0.007612g of NH₄SCN.

Procedure for assay:

Assay of theophylline :

- Dilute a solution equivalent to 0.25 g of aminophylline up to 40.0 ml using distilled water.
- Warm gently on a water bath until complete dissolution is effected.
- Add 0.1 M AgNO₃ (20.0 ml), boil for 15 minutes and cool to 5°C – 10°C, filter and wash.
- Wash the precipitate with nitric acid (HNO₃) and add 3.0 ml of an excess of HNO₃ followed by ferric ammonium sulfate (2.0 ml) and titrate against 0.1 M NH₄SCN.
- Each ml of 0.1 M AgNO₃ is equivalent to 0.01802 g of theophylline.

Assay of ethylene diamine:

- Dilute a solution equivalent to 0.25 g of aminophylline up to 30.0 ml using distilled water.
- Titrate the resulting solution with 0.1 M HCl using methyl red as indicator until the color changes from yellow to red.

Category:

- Bronchodialator, antiasthmatic.

Standards:

Aminophylline contains an equivalent of NLT 84% and NMT 87.4% of theophylline and an equivalent of NLT 35.5% and NMT 50% of ethylene diamine.

## 10

Assay of Atropine Sulphate

Aim: To perform the assay of the alkaloid, atropine as its sulphate salt.

Apparatus: Conical flask, beaker, burette, burette stand, glazed tile.

Chemicals:

Chemicals & Reagents	Quantity Taken
Atropine sulphate	0.5 g
Glacial acetic acid	50 ml
Potassium hydrogen phthalate	0.3 g
Crystal violet	Few drops
Perchloric acid	*q.s.*

Principle:

Atropine sulphate is freely soluble in water but its assay in aqueous medium does not give sharp end point due to its weak alkalinity. Hence it is assayed by using non aqueous titrimetry using perchloric acid in the presence of a protophilic solvent like glacial acetic acid.

Atropine belongs to the tropane class of alkaloids and is isolated from the plants *Atropa belladonna*, *Datura stramonium* and *Hyoscyamus niger*.

In nonaqueous titration, nonaqueous solvents such as CH_3COOH (100%), acetone and weak organic bases, like tetrabutylammonium hydroxide in methanol, potassium, sodium, or lithium methoxide in toluene-methanol, are used for titration of weekly basic and weekly acidic substances respectively.

Weekly acidic and basic substances do not give sharp end points in aqueous titrimetry due to its low dissociation constant. The amphiprotic solvent, like acetic acid, which is used as a solvent in non aqueous titrations of basic substances functions both as an acid and as a base.

When it is mixed with perchloric acid, it acts as a base and accepts proton. In the presence of a base like pyridine, it acts as an acid by donating protons. Thus it increases the acidity or basicity of the weekly acidic or basic substances hence enhances the attainment of sharp endpoint.

In the preparation of perchloric acid (0.1 M) solution, acetic anhydride is added to absorb the moisture in the perchloric acid and acetic acid, rendering it virtually anhydrous. Though excess of acetic anhydride is not always disadvantageous, care must be taken to avoid excess when primary and secondary amines have to be titrated. It readily acylate them to give neutral products and reduce its basicity. The perchloric acid must be well diluted with acetic acid before adding the acetic anhydride. If it is not done it may lead to explosive acetyl perchlorate.

Potassium hydrogen phthalate is used as a primary standard for standardization of perchloric acid.

Reaction:

$$HClO_4 + CH_3COOH \longrightarrow CH_3COOH_2^{(+)} + ClO_4^{(-)}$$

$$CH_3COO^{(-)} + CH_3COOH_2^{(+)} \longrightarrow CH_3COOH + CH_3COOH$$

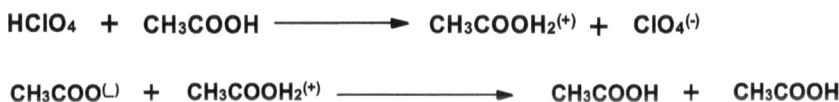

Preparation of 0.1 M perchloric acid:

- Mix an accurately measured volume of Perchloric acid (8.5 ml) with anhydrous glacial acid (500.0 ml).
- Then add 25.0 ml of acetic anhydride.
- Cool the mixture to room temperature and make up the volume to 1000 ml using anhydrous glacial acetic acid.
- Allow the solution to stand for over night.

Standardisation of 0.1 M perchloric acid:

- Accurately weigh 0.1 g of potassium hydrogen phthalate and dissolve it in 10.0 ml of glacial acetic acid.

- Add a few drops of crystal violet solution and titrate with 0.1 M perchloric acid until the color changes from violet to emerald green.
- Perform a blank determination and make necessary corrections.
- Equivalent factor: Each ml of 0.1 M perchloric acid = 0.02042 g of potassium hydrogen phthalate.

Procedure:

- Weigh accurately about 0.5 g of atropine and dissolve in 30.0 ml of anhydrous glacial acetic acid.
- Titrate against standard perchloric acid using crystal violet as indicator until the color changes from bluish violet to emerald green.
- Perform a blank determination and incorporate any necessary corrections.
- Each ml 0.1 M perchloric acid is equivalent to 0.06770 g of atropine sulphate.

Standards:

Contains not less than 99 % and not more than 101 % of atropine sulphate calculated with reference to anhydrous substance.

Category:

- Physiologically atropine has anticholinergic activity.
- It is used as an antidote to cholinesterase poisoning.

References:

1. Indian Pharmacopoeia 1996 Vol 1 Ministry of Health and Family Welfare, Govt. of India pp. 76

11

Assay of Ephedrine Hydrochloride

Aim: To perform the assay of the alkaloid, ephedrine hydrochloride.

Apparatus: Conical flask, beaker, burette stand, glazed tile.

Chemicals:

Chemicals & Reagents	Quantity Taken
Ephedrine hydrochloride	0.17 g
Mercuric acetate solution	10.0 ml
Actone	50.0 ml
Potassium hydrogen phthalate	0.3 g
Crystal violet	Few drops
Perchloric acid	*q.s.*

Principle:

Ephedrine is chemically (1R, 1S)-2- methylamino-1-phenylpropan-1-ol. It is an alkaloid obtained naturally from the genus *Ephedra* and also prepared synthetically. Physiologically it has sympathomimetic and bronchodilator activities. The assay of ephedrine is carried out by nonaqueous titrimetry using perchloric acid as they are weak bases. The end point is determined using crystal violet as indicator and the color change is from bluish violet to emerald green.

253

In nonaqueous titration, organic solvents such as CH_3COOH (100%), acetone and weak organic bases, like tetrabutylammonium hydroxide in methanol, potassium, sodium, or lithium methoxide in toluene-methanol, are used for titration of weekly basic and weekly acidic substances respectively.

Weekly acidic and basic substances do not give sharp end points in aqueous titrimetry due to its low dissociation constant. The amphiprotic solvent, like acetic acid, which is used as a solvent in non aqueous titrations of basic substances functions both as an acid and as a base.

When it is mixed with perchloric acid, it acts as a base and accepts proton. In the presence of a base like pyridine, it acts as an acid by donating protons. Thus it increases the acidity or basicity of the weekly acidic or basic substances hence enhances the attainment of sharp endpoint.

In the preparation of perchloric acid (0.1 M) solution, acetic anhydride is added to absorb the moisture in the perchloric acid and acetic acid rendering it virtually anhydrous. Though excess of acetic anhydride is not always disadvantageous, care must be taken to avoid excess when primary and secondary amines have to be titrated. It readily acylate them to give neutral products reduce its basicity. The perchloric acid must be well diluted with acetic acid before adding the acetic anhydride. If it is not done it may lead to explosive acetyl perchlorate.

Potassium hydrogen phthalate is used as a primary standard for standardization of perchloric acid.

Reaction:

$$HClO_4 + CH_3COOH \longrightarrow CH_3COOH_2^{(+)} + ClO_4^{(-)}$$

$$CH_3COO^{(-)} + CH_3COOH_2^{(+)} \longrightarrow CH_3COOH + CH_3COOH$$

Preparation of 0.1 M perchloric acid:

- Mix an accurately measured volume of perchloric acid (8.5 ml) with anhydrous glacial acid (500.0 ml).
- Then add 25.0 ml of acetic anhydride.

- Cool the mixture to room temperature and make up the volume to 1000 ml using anhydrous glacial acetic acid.
- Allow the solution to stand over night.

Standardisation of 0.1 M perchloric acid:

- Accurately weigh 0.1 g of potassium hydrogen phthalate and dissolve it in 10.0 ml of glacial acetic acid.
- Add a few drops of crystal violet solution and titrate with 0.1 M perchloric acid until the color changes from violet to emerald green.
- Perform a blank determination and make necessary corrections.
- Equivalent factor: Each ml of 0.1 M perchloric acid = 0.02042 g of potassium hydrogen phthalate.

Procedure:

- Weigh accurately about 0.17 g of ephedrine hydrochloride and dissolve in 10.0 ml of mercuric acetate solution warming gently. Add 50.0 ml of acetone.
- Titrate the resulting solution against standard perchloric acid using crystal violet as indicator until the appearance of emerald green color.
- Perform a blank determination and make any necessary corrections.
- Each ml 0.1 M perchloric acid is equivalent to 0.02017g of ephedrine HCl.

Standards:

Contains not less than 98.5% and not more than 101% of ephedrine hydrochloride calculated with reference to anhydrous substance.

Category:

Sympathomimetic, bronchodilator.

References:

1. Indian Pharmacopoeia 1996 vol 1 Ministry of Health and Family Welfare, Govt. of India pp. 283

## 12

Assay of Caffeine

Aim: To perform the assay of the alkaloid, caffeine.

Apparatus: Conical flask, beaker, burette, burette stand, glazed tile.

Chemicals:

Chemicals & Reagents	Quantity Taken
Caffeine	0.18 g
Glacial acetic acid solution	50.0 ml
Potassium hydrogen phthalate	0.3 g
Crystal violet	Few drops
Perchloric acid	*q.s.*

Principle:

Caffeine ($C_8H_{10}N_4O_2$) is chemically 1,3,7-trimethylxanthine or 3,7-dihydro-1,3,7-trimethyl-1-H-purine-2,6-dione. It is naturally produced by several plants, including coffee beans, guarana, yerba mate, cacao beans and tea. Caffeine is assayed by non aqueous titrimetry using perchloric acid. Crystal violet is used as an indicator and the color change at the endpoint is from bluish violet to emerald green.

In nonaqueous titration, nonaqueous solvents such as CH_3COOH (100%), acetone and weak organic bases, like tetrabutylammonium hydroxide in methanol, potassium, sodium, or lithium methoxide in toluene-methanol, are used for titration of weekly basic and weekly acidic substances respectively.

Weekly acidic and basic substances do not give sharp end points in aqueous titrimetry due to its low dissociation constant. The amphiprotic solvent, like acetic acid, which is used as a solvent in nonaqueous titrations of basic substances functions both as an acid and as a base.

When it is mixed with perchloric acid, it acts as a base and accepts proton. In the presence of a base like pyridine, it acts as an acid by donating protons. Thus it increases the acidity or basicity of the weekly acidic or basic substances hence enhances the attainment of sharp endpoint.

In the preparation of perchloric acid (0.1 M) solution, acetic anhydride is added to absorb the moisture in the perchloric acid and acetic acid rendering it virtually anhydrous. Though excess of acetic anhydride is not always disadvantageous, care must be taken to avoid excess when primary and secondary amines have to be titrated. It readily acylate them to give neutral products and reduce its basicity. The perchloric acid must be well diluted with acetic acid before adding the acetic anhydride. If it is not done it may lead to explosive acetyl perchlorate.

Potassium hydrogen phthalate is used as a primary standard for standardization of perchloric acid.

Reaction:

$$HClO_4 + CH_3COOH \longrightarrow CH_3COOH_2^{(+)} + ClO_4^{(-)}$$

$$CH_3COO^{(-)} + CH_3COOH_2^{(+)} \longrightarrow CH_3COOH + CH_3COOH$$

Preparation of 0.1 M perchloric acid:

- Mix an accurately measured volume of perchloric acid (8.5 ml) with anhydrous glacial acid (500.0 ml).
- Then add 25.0 ml of acetic anhydride.
- Cool the mixture to room temperature and make up the volume to 1000 ml using anhydrous glacial acetic acid.
- Allow the solution to stand for over night.

Standardisation of 0.1 M perchloric acid:

- Accurately weigh 0.1 g of potassium hydrogen phthalate and dissolve it in 10.0 ml of glacial acetic acid.
- Add a few drops of crystal violet solution and titrate with 0.1 M perchloric acid until the color changes from violet to emerald green.

- Perform a blank determination and make necessary corrections.
- Equivalent factor: Each ml of 0.1 M perchloric acid = 0.02042 g of potassium hydrogen phthalate.

Procedure:

- Weigh accurately about 0.18 g of caffeine and dissolve with warming in 5.0 ml of anhydrous glacial acetic acid. For caffeine hydrate, use material previously dried at $100^{\circ}C$ to $105^{\circ}C$.
- Cool and add 10.0 ml of acetic anhydride and 20.0 ml of toluene.
- Titrate against standard perchloric acid using crystal violet as indicator until the color changes from bluish violet to emerald green.
- Perform a blank determination and make any necessary corrections.
- Each 0.1 M perchloric acid is equivalent to 0.01942 g of caffeine.

Standards:

Contains not less than 98.5% and not more than 101.5% of caffeine calculated with reference to dried substance.

Category:

CNS stimulant.

References:

1. Indian Pharmacopoeia 1996 Vol 1 Ministry of Health and Family Welfare, Govt. of India pp. 121

13

Assay of Adrenaline

Aim: To perform the assay of the alkaloid, adrenaline.

Apparatus: Conical flask, beaker, burette, burette stand, glazed tile.

Chemicals:

Chemicals & Reagents	Quantity Taken
Adrenaline	0.3 g
Glacial acetic acid solution	50.0 ml
Potassium hydrogen phthalate	0.3 g
Crystal violet	Few drops
Perchloric acid	*q.s.*

Principle:

Adrenaline is chemically -1-(3,4-Dihydroxyphenyl)-2-methyl-amino ethanol. Physiologically it has sympathomimetic and bronchodialator activities.

The assay of adrenaline is carried out by non aqueous titrimetry using perchloric acid. The end point is determined using crystal voilet as indicator and the color change is from blue to emerald green.

In nonaqueous titration, nonaqueous solvents such as CH_3COOH (100%), acetone and weak organic bases, like tetrabutylammonium hydroxide in methanol, potassium, sodium, or lithium methoxide in toluene-methanol, are used for titration of weekly basic and weekly acidic substances respectively.

Weekly acidic and basic substances do not give sharp end points in aqueous titrimetry due to its low dissociation constant. The amphiprotic solvent, like acetic acid, which is used as a solvent in non aqueous titrations of basic substances functions both as an acid and as a base. When it is mixed with perchloric acid, it acts as a base and accepts

259

proton. In the presence of a base like pyridine, it acts as an acid by donating protons. Thus it increases the acidity or basicity of the weekly acidic or basic substances hence enhances the attainment of sharp endpoint.

In the preparation of perchloric acid (0.1 M) solution, acetic anhydride is added to absorb the moisture in the perchloric acid and acetic acid rendering it virtually anhydrous. Though excess of acetic anhydride is not always disadvantageous, care must be taken to avoid excess when primary and secondary amines have to be titrated. It readily acylate them to give neutral products and reduce its basicity. The perchloric acid must be well diluted with acetic acid before adding the acetic anhydride. If it is not done it may lead to explosive acetyl perchlorate.

Potassium hydrogen phthalate is used as a primary standard for standardization of perchloric acid.

Reaction:

$$HClO_4 + CH_3COOH \longrightarrow CH_3COOH_2^{(+)} + ClO_4^{(-)}$$

$$CH_3COO^{(-)} + CH_3COOH_2^{(+)} \longrightarrow CH_3COOH + CH_3COOH$$

Preparation of 0.1 M perchloric acid:
- Mix an accurately measured volume of perchloric acid (8.5 ml) with anhydrous glacial acid (500.0 ml).
- Then add 25.0 ml of acetic anhydride.
- Cool the mixture to room temperature and make up the volume to 1000 ml using anhydrous glacial acetic acid.
- Allow the solution to stand for over night.

Standardisation of 0.1 M perchloric acid:
- Accurately weigh 0.1 g of potassium hydrogen phthalate and dissolve it in 10.0 ml of glacial acetic acid.
- Add a few drops of crystal violet solution and titrate with 0.1 M perchloric acid until the color changes from violet to emerald green.

- Perform a blank determination and make necessary corrections.
- Equivalent factor: Each ml of 0.1 M perchloric acid = 0.02042 g of potassium hydrogen phthalate.

Procedure:

- Weigh accurately about 0.3 g of adrenaline and dissolve in 50.0 ml of anhydrous glacial acetic acid with warming if necessary to effect solution.
- Titrate against standardized perchloric acid using crystal violet as indicator until the color changes from bluish violet to emerald green.
- Perform a blank determination and make any necessary corrections.
- Each ml 0.1M perchloric acid is equivalent to 0.01832 g of adrenaline.

Standards:

Contains not less than 98.5% and not more than 101% of adrenaline calculated with reference to dried substance.

Category:

Sympathomimetic.

References:

1. Indian Pharmacopoeia 1996 Vol 1 Ministry of Health and Family Welfare, Govt. of India pp. 22

## 14

Identification of Durgs

Sample-1 Quinine Sulphate

Description: White needle like crystal or crystalline powder.

Solubility: Freely soluble in a mixture of 2 volumes of chloroform ($CHCl_3$) and one volume ethanol.

Sparingly soluble in boiling water and ethanol.

Soluble in water, practically insoluble in ether.

Tests:

To 0.5ml of 0.1% $^W/_V$ solution add 0.2 ml of bromine solution and 1ml of dilute nitric acid (HNO_3) an emerald green color was produced.

To 0.5% $^W/_V$ solution, an equal volume of dilute sulphuric acid (H_2SO_4) was added, a light blue fluorescence was produced.

To 5ml of 1% $^W/_V$ solution, 1ml of $AgNO_3$ solution was added, stirred after a short interval, white precipitate soluble in HNO_3 was produced.

0.75 mg of substance was dissolved in 5 ml of water. To it add 1ml of dilute HCl and 1 ml of Barium chloride ($BaCl_2$) solution. A white precipitate was observed.

0.50 mg of substance was dissolved in 5 ml of water, add lead acetate 2 ml, a white precipitate was observed which is soluble in ammonium acetate and NaOH.

Sample-2 Calcium Gluconate

Description: White crystalline powder.

Solubility: Sparingly soluble in water, freely soluble in boiling water, and insoluble in ethanol.

Test-1: To 10 ml of 3% $^W/_V$ solution 0.05ml of ferric chloride ($FeCl_3$) solution was added, yellow color was observed.

Test-2: 20 mg of substance was dissolved in glacial CH_3COOH and 0.05 ml of ferro cyanide, 450 mg of ammonium chloride (NH_4Cl) were added. White crystalline ppt was observed.

Test-3: To 5 ml of 0.4 $^W/_V$ solution, 0.2 ml of 2% $^W/_V$ of ammonium oxalate was added. Precipitate was obtained which was only sparingly soluble in dil CH_3COOH but completely soluble in HCl.

Sample-3: Sulphaguanidine

Discription: White crystalline powder.

Solubility: Freely soluble in acetone, sparingly soluble in ethyl alcohol, slightly soluble in chloroform ($CHCl_3$) and ether, practically insoluble in water, dissolves in dilute alkali solution.

Tests:

0.1 g of substance was dissolved in 2 ml of HCl. To this 0.2 ml of $NaNO_2$ solution was added and cooled to $0°C$-$5°C$. After 12 minutes add this solution to 1 ml of ice cold β-naphthol. An intense orange to red color dye is observed.

Sample-4 Dapsone

Description: White crystalline powder.

Solubility: Freely soluble in ethyl alcohol, acetone.

Very slightly soluble in water and soluble in dilute mineral acid.

Tests:

0.1g substance was dissolved in 2.0 ml of 2 M HCl and 0.2 ml of $NaNO_2$ solution was added. After 1 or 2 minutes the solution was added to 1ml of β-napthol. An orange/red dye was produced.

Melts between $175°C$-$185°C$.

Sample-5 Amoxycilin Trihydrate

Description: White crystalline powder.

Solubility: Slightly soluble in water, methanol, ethanol.

Insoluble in chloroform, ether and fixed oils.

Tests:

1. To drug solution, 2 ml of mixture of potassium tartarate and 3 ml of water was added. A magenta color was observed.

2. To the drug solution, formaldehyde (HCHO) and H_2SO_4 were added in small quantities. Initially solution was colorless but turns dark yellow after mixing.

Part E

Analysis of Oils and Fats

Determination of Acid Value of given Oil

Aim: To determine the acid value of a given sample of Arachis oil.

Apparatus: Conical flask, round bottomed flask, beaker, burette, pipette, weighing bottle, fractional weights.

Chemicals: Sodium carbonate, potassium hydroxide, hydrochloric acid, methyl orange, phenolphthalein, arachis oil.

Principle:

Acid value is defined as the number of milligrams of potassium hydroxide (KOH) required to neutralize the free fatty acid present in 1g of oil or fat. The acid value is a measure of the amount of carboxylic acid groups in a chemical compound, such as a fatty acid, or in a mixture of compounds. As oil-fats rancify, triglycerides are converted into fatty acids and glycerol, causing an increase in the magnitude of its acid value.

Procedure:

1. Accurately weigh 1g of given oil sample into a dry conical flask.
2. Add 5 ml of each of ether and alcohol which are previously neutralized with 0.1N KOH solution to phenolphthalein.
3. Heat the mixture on a water bath gently if necessary until the oily substance dissolves completely.
4. Titrate the resulting solution with standard 0.1 N KOH solution with constant shaking until a pink color appears and persists for 15 seconds.
5. Note down the number of ml of KOH required and calculate the acid value using the following formula.

$$\text{ACID VALUE} = A \times 0.00561 \times \frac{1000}{W} \text{ where}$$

A = Number of ml of 0.1 N KOH consumed

W = The weight in grams of oil or fat

Standard values:

Arachis oil	:	NMT 0.5
Linseed oil	:	NMT 4
Sesame oil	:	NMT 2
Cod-liver oil	:	NMT 2

References:

1. Indian Pharmacopoeia 1996 vol 2 (p-z) Appendix 3.2 Ministry of Health and Family Welfare, Govt. of India pp A-51.

2

Determination of Saponification Value of given Oil

Aim: To determine saponification value of the given oil.

Apparatus: Conical flask, round bottomed flask, weighing balance, fractional weight, water, reflux condenser.

Chemicals: Potassium hydroxide, oil, alcohol, hydrochloric acid,

Principle:

Saponification value is the number of milligrams of KOH required to neutralize the free fatty acid and to saponify the esters in 1g of given oil or fat. Oils and fats are glyceryl esters of fatty acids. It is a measure of the average molecular weight (or chain length) of all the fatty acids present. As most of the mass of the fat is in the free fatty acids, it allows for comparison of the average fatty acid chain length.

Procedure:

1. Accurately weigh 0.5 g of given oil into a round bottomed flask.
2. Add 10 ml of alcoholic KOH solution and heat the mixture under reflux for 1 hour, by frequently shaking the contents of the flask.
3. Cool the reaction mixture and titrate the resulting solution with standard HCl solution.
4. Note down the volume of HCl consumed as A.
5. Perform a blank determination omitting the substance being examined and note down the volume as B.
6. Calculate the saponification value using the formula.

$$\text{SAPONIFICATION VALUE} \ = \ (\text{B-A}) \times \frac{28.05}{\text{W}}$$

Where,

A = Number of ml of 0.1N KOH consumed by test

B = Number of ml of 0.1N KOH consumed by blank

W = The weight in grams of oil or fat.

Standard values: Arachis oil   :   185-195

Linseed oil   :   188-195

Sesame oil   :   188-195

Cod-liver oil   :   180-190

Reference:

1. Indian Pharmacopoeia -1996 vol 2 (p-z) Appendix-3.2Ministry of Health and Family Welfare, Govt. of India pp A-51

3

Determination of Ester Value of given Oil

Aim: To determine ester value of the given oil.

Apparatus: Conical flask, round bottomed flask, weighing balance, fractional weight, water, reflux condenser.

Chemicals: Sodium carbonate, potassium hydroxide, hydrochloric acid, methyl orange, phenolphthalein, arachis oil.

Principle:

Ester value is defined as the no. of milligrams of KOH required to saponify the esters present in the given oil sample.

Ester value = Saponification value – Acid value

References:

1. Indian Pharmacopoeia 1996 vol 2 (p-z) Appendix 3.2 Ministry of Health and Family Welfare, Govt. of India pp A-51

4

Determination of
Iodine Value of given Oil

Aim: To determine iodine value of the given oil sample.

Apparatus: Iodine flask, pipette, burette, measuring cylinder, fractional weights, weighing box, spatula.

Chemicals: Arachis oil, carbon tetrachloride, pyridine bromide solution, potassium iodide, sodium thiosulphate, starch indicator.

Principle:

Iodine value is the weight of iodine absorbed by 100 parts by weight of an oil or fat. Iodine gets directly added upon to double bonds. Thus Iodine value indicates the amount of unsaturation present in oil or fat. Oils have higher iodine value as they are unsaturated whereas fats has lesser iodine value due to their saturated property. Non drying oils have iodine value of 80-100 where as drying oils have higher iodine value. Iodine value can be determined by three methods.

1. Iodine mono chloride method (Wijs method)
2. Iodine mono bromide method (Hanus method)
3. Pryidine bromide method

   Pyridine bromide method makes use of an additive compound of pyridine, bromine, sulphuric acid. These reagents form additive compounds with unsaturation present in given oil or fat without having any side reactions like oxidation or substitution. The excess of pyridine bromide added will react with potassium iodide and liberates equivalent amount of iodine. The iodine formed will be titrated with sodium thiosulphate solution with starch as an indicator.

Procedure:

1. Place an accurately weighed quantity of oil (0.5 g) in a dry iodine flask.

2. Add 10 ml carbon tetrachloride (CCl_4) to dissolve the oil.

3. To this solution add 25 ml of pyridine bromide solution and allow to stand for half an hour in dark.

4. As the temperature is in the range of 15°C-25°C, add 15 ml potassium iodide (KI) solution taken on a cuptap apparatus by removing the stopper carefully. Rinse both the stopper and sides of iodine flask with 50 ml of water.

5. Titrate the reaction mixture with standard sodium thiosulphate ($Na_2S_2O_3$) solution using starch solution as indicator which is added as end point approaches.

6. Note down the volume of sodium thiosulphate rundown as A.

7. Perform a blank determination and note down the volume consumed as B.

8. Calculate the Iodine value using formula.

$$\text{IODINE VALUE} = 1.26 \times \frac{(B-A)}{W}$$

Where,

A = Number of ml of 0.1 N $Na_2S_2O_3$ consumed by test

B = Number of ml of 0.1 N $Na_2S_2O_3$ consumed by blank

W = The weight in grams of oil or fat.

Standard values:	Arachis oil	:	85-105
	Linseed oil	:	160-200
	Sesame oil	:	103-116
	Codliver oil	:	145-180

References:

1. Indian Pharmacopoeia -1996 vol 2 (p-z) Appendix – 3.2 Ministry of Health and Family Welfare, Govt. of India pp A-51

5

Determination of Unsaponifiable Matter Present in the given Oil

Aim: To determine the unsaponifiable matter present in the given oil.

Apparatus: Round bottomed flask (RBF), beaker, separating funnel, water bath, pipette etc.

Chemicals: Potassium hydroxide, oil, methanol, ether, phenolphthalein.

Principle:

The unsaponifiable matter consists of substance present in the oil or fat which are not saponifiable by alkali hydroxide and are determined by extracting with an organic solvent of a solution of saponified substance being examined.

Procedure:

1. Accurately weigh 2.5 g of oil sample into a 250 ml round bottomed flask fitted with a reflux condenser.

2. Add a solution of 2 g of KOH in 40 ml of methanol. Heat the reaction mixture on a water bath for 1 hour.

3. Transfer the contents of flask was into a separating funnel with the aid of 50 ml of hot water.

4. While the liquid is still warm shake very carefully with 50 ml of peroxide-free ether (thrice).

5. Combine the ether extracts in a second separating funnel containing 40 ml water, swirl the mixture gently for few minutes and allow to separate.

6. Reject the lower aqueous layer and wash the ether extract with 2 quantities each of 40 ml of water and 3 quantities each of 40 ml of 3% $^W/_V$ solution of KOH, each treatment being followed by a washing with 40 ml water.

7. Finally wash the ether layer with successive quantities each of 40 ml of water until the aqueous layer is not alkaline to phenolphthalein.

8. Transfer the ether extracts into previously weighed flask and wash the separating funnel with peroxide free ether and add to the previously weighed flask.

9. Distil off the ether and add 6.0 ml of acetone to the residue and further distil off.

10. Dry the reaction vessel at 100°C-105°C for 30 minutes and cool in a desiccator. Weigh the residue and calculate the unsaponifiable matter as % $^W/_W$.

Standard value: Not more than 1.5%.

6

Determination of Peroxide Value of the given Oil

Aim: To determine the peroxide value of the given oil.

Apparatus: Iodine flask, burette, pipette, weighing capsule.

Chemicals: Oil, glacial acetic acid, potassium iodide, chloroform, sodium thiosulphate.

Principle:

Peroxide value is number of milliequivalents of active oxygen that is expressed as amount of peroxide contained in 100 g of oil or fat substance.

Procedure:

1. Accurately weigh 2.5 g of the oil sample in 250 ml iodine flask.
2. Add 15ml of a mixture of 3 volume in glacial acetic acid (9 ml) and 2 volumes of chloroform (6 ml). Shake or swirl the iodine flask until the oil dissolves.
3. Add 0.5 ml of saturated potassium iodide and titrate the resulting mixture with sodium thiosulphate solution until the yellow color disappears.
4. Now add 0.5 ml of starch solution and continue titration until the blue color disappears.
5. Note down the volume consumed as A.
6. Perform a blank determination and note down the volume consumed as B.
7. Calculate the peroxide value as

$$\text{Peroxide value} = \frac{10(A-B)}{W}$$

Where,

A = Number of ml of 0.1N $Na_2S_2O_3$ consumed by test

B = Number of ml of 0.1N $Na_2S_2O_3$ consumed by blank

W = The weight in grams of oil or fat.

Reference:

1. Indian Pharmacopoeia -1996 vol 2 (p-z) Appendix -3.2 Ministry of Health and Family Welfare, Govt. of India pp A-51.

Part F

Interpretation of
Different Spectra

Woodward-Fieser Rules for calculating λ_{max} Values for Dienes:

Increase in the length of conjugated system increases the λ_{max} and also ε_{max}.

Conjugated polyene system with more than 5 double bonds will be coloured and has λ_{max} in visible regions *i.e.,* above 400 nm.

Alkyl group on double bond causes bathochromic shift.

The various possible conjugated dienes are as follows

1. Alicyclic dienes: Base unit butadiene.
2. Homoannular conjugated dienes (Homodienes): Conjugated double bonds within the same ring system.

3. Hetero annular conjugated dienes: Conjugated double bonds not in the same ring.

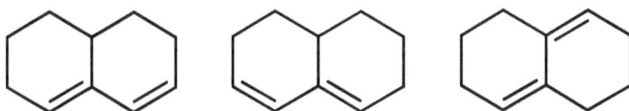

4. Exocyclic and endocyclic conjugated double bonds: Exocyclic double bond is a double bond part of the conjugated system formed by any carbon atom of any ring but present outside ring. Endocyclic double bond is present inside the ring.

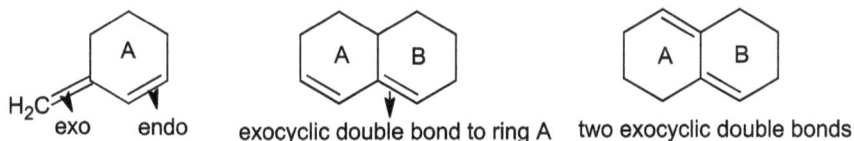

A

H₂C
exo   endo   exocyclic double bond to ring A   two exocyclic double bonds

A   B   A   B

According to these rules, each type of diene has a certain fixed basic value and the value of λ_{max} **depends on :**

(i)   The no. of alkyl groups and the ring residues on the double bond.
(ii) The no. of double bonds which extend conjugation.
(iii) The presence of polar groups such as –Cl, –Br, –OR, –SR etc.

281

The various rules for calculating the absorption maximum in case of dienes and trienes are summarized in the table:

Conjugated dienes and trienes Solvent-Ethanol

Parent value for butadiene system or a cyclic conjugated diene	217 nm
Acyclic triene	245 nm
Homoannular diene	253 nm
Heteroannular diene	215 nm
Increments for each substituent	
Alkyl substituent or Ring residues	5 nm
Exocyclic double bonds	5 nm
Double bond extending conjugation	30 nm
Auxochrome	
-OR	+6 nm
-Cl, -Br	+5 nm
-SR	+30 nm
NR$_2$	+60 nm
OCOCH$_3$	0 nm

1. **To determine the λ_{max} of following molecules using Woodward Fieser Rule**

   1)

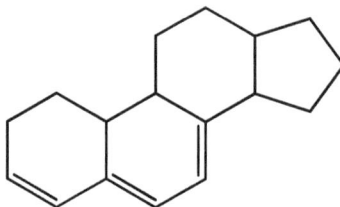

Base value for Homoannular diene		= 253 nm
Increments for:		
4 Alkyl ring residues	4×5	= 20 nm
1 Double bond extending conjugation	1×30	= 30 nm
2 Exocyclic double bonds	2×5	= 10 nm
TOTAL		**= 313 nm**

2)

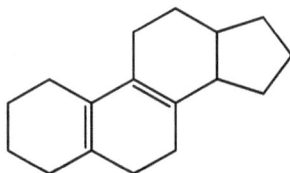

Base value for Homoannular diene	= 253 nm
Increments for:	
6 Alkyl Ring residues	$6 \times 5 = 30$ nm
TOTAL	**= 283 nm**

3)

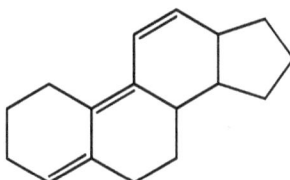

Base value for Heteroannular diene	= 214 nm
Increments for:	
5 Alkyl ring residues	$5 \times 5 = 25$ nm
1 Double bond extending conjugation	$1 \times 30 = 30$ nm
3 Exocyclic double bonds	$3 \times 5 = 15$ nm
TOTAL	**= 284 nm**

4)

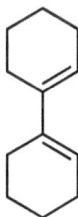

Base value for Heteroannular diene	= 215 nm
Increments for:	
4 Alkyl Ring residues	$4 \times 5 = 20$ nm
TOTAL	**= 235 nm**

5)

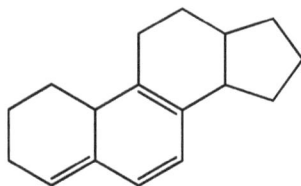

Base value for Homoannular diene		= 253 nm
Increments for:		
5 Alkyl ring residues	$5 \times 5 = 25$ nm	
1 Double bond extending conjugation	$1 \times 30 = 30$ nm	
1 Exocyclic double bond	$1 \times 5 = 5$ nm	
TOTAL		**= 313 nm**

Woodward-Fieser Rules for calculating λ_{max} in α, β-unsaturated carbonyl compounds:

(a) The base value of α, β-unsaturated ketone is taken as 215 nm. The α, β-unsaturated ketone may be a cyclic or six membered.

For a compound, = CH-COX, base value is 215 nm, if X is an alkyl group.

If X = H, base value becomes 207 nm. The basic value is 193 nm if X is OH or OR.

(b) If the double bond and the carbonyl group are contained in a five membered ring (cyclopentanone), then for such an α, β-unsaturated ketone, the basic value becomes 202 nm.

The structural increments for estimating λ_{max} for a given α, β-unsaturated carbonyl compounds is as follows:

- For each exocyclic double bond   +5 nm

- For each endocyclic double bond in five or
  seven membered ring except cyclo-pent-2 enone   +5 nm

- For each alkyl substituent or ring residue at the

  α position   +10 nm

  β position   +12 nm

  γ position or higher position   +18 nm

- For each double bond extending conjugation   +30 nm
- For a homoanular conjugated diene   +39 nm

Increments for various auxochromes in the various α, β, γ etc. Positions are given in the following table:

Chromophore increments in carbonyl compounds

Chromophore	Increment in nm for position w.r.to the carbonyl group			
	α	β	γ	Higher groups
-OH	+35	+30	---	+50
-OAc	+6	+6	+6	+6
-Cl	+15	+12	---	---
-Br	+25	+35	---	---
-OR	+35	+30	+17	+31
-SR	---	+85	---	---
-NR$_2$	---	+95	---	---

Determine the λ_{max} of following molecules using Woodward Fischer Rule

1)

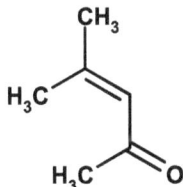

Base value for α,β unsaturated ketone   = 215 nm

Increments for:

2 β- alkyl substituents $2 \times 12 = 24$ nm

TOTAL   **= 239 nm**

2)

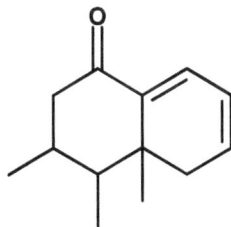

Base value for α, β unsaturated ketone   = 215 nm

Increments for:

1 α-Ring residue	1 × 10 = 10 nm
1 γ'- Ring residue	1 × 18 = 18 nm
1 Exocyclic double bond	1 × 5 = 5 nm
Homoannular diene	= 39 nm
TOTAL	**= 287 nm**

3)

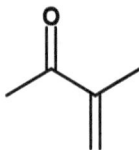

Base value for α, β unsaturated ketone   = 215 nm

Increments for:

1 α-Ring residue	1 × 10 = 10 nm
TOTAL	**= 225 nm**

4)

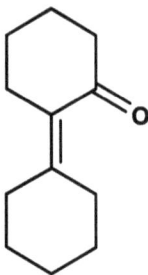

Base value for α, β unsaturated ketone   = 215 nm

Increments for:

1 α-Ring residue	1 × 10 = 10 nm
2 β- Ring residue	2 × 12 = 24 nm
2 Exocyclic double bonds	2 × 5 = 10 nm
TOTAL	**= 259 nm**

5)

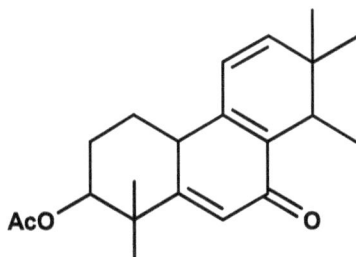

Base value for α, β unsaturated ketone	= 215 nm
Increments for:	
1 α-Ring residue	1 × 10 = 10 nm
1 β-Ring residue	1 × 12 = 12 nm
1 γ-Ring residue	1 × 18 = 18 nm
1 Double bond extending conjugation	1 × 30 = 30 nm
Homoannular diene	= 39 nm
TOTAL	**= 324 nm**

6)

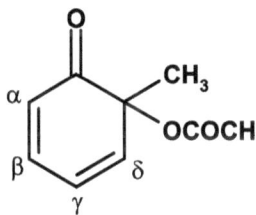

Base value for six membered enone	= 215 nm
Increments for:	
1 Double bond extending conjugation	1 × 30 = 30 nm
1 δ- Ring residue	1 × 18 = 18 nm
Homoannular diene	= 39 nm
TOTAL	**= 302 nm**

7)

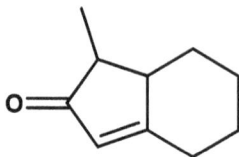

Base value for five membered enone	= 202 nm
Increments for:	
2 β- Ring residue	2 × 12 = 24 nm
1 Exocyclic double bond	1 × 5 = 5 nm
TOTAL	**= 231 nm**

8)

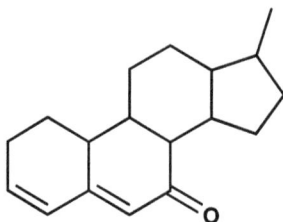

Base value for six membered enone	= 215 nm
Increments for:	
1 Double bond extending conjugation	1 × 30 = 30 nm
1 β- Ring residue	1 × 12 = 12 nm
1 δ- Ring residue	1 × 18 = 18 nm
1 Exocyclic double bond	1 × 5 = 5 nm
TOTAL	**= 280 nm**

9)

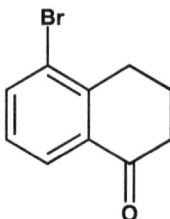

Base value for parent chromophore	=	246 nm
O-Ring residue	=	3 nm
m-Br	=	2 nm
TOTAL	=	**251 nm**

Derivatives of acyl compounds:

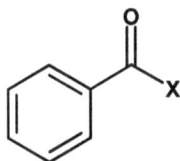

The base value is 246 nm. If X, is an alkyl group or alicyclic residue.

If X is hydrogen atom, the base value becomes 250 nm and the base value is 230 nm if x is OH or OR. The structural increments in further substitution on the aromatic ring in the ortho, meta and para positions are given in the table:

Auxochrome acting as a substituent

Auxochrome	Increments in nm position of substituent		
	Ortho	Meta	Para
Alkyl	+3	+3	+10
OH,OR	+7	+7	+25
Cl	0	0	+10
Br	+2	+2	+15
NH_2	+13	+13	+58
NHAc	+20	+20	+45
NR_2	+20	+20	+85
O^-	+11	+20	+75

1)

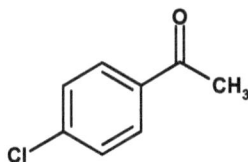

Base value	=	246 nm
Increments for:		
Cl-substitution at para position	=	10 nm
TOTAL	=	**256 nm**

2)

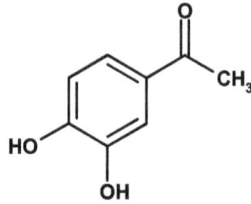

Base value	=	246 nm
Increments for:		
OH-substitution at meta position	=	7 nm
OH-substitution at para position	=	25 nm
TOTAL	=	**278 nm**

3)

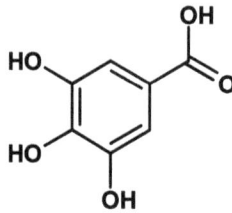

Base value for Parent chromophore	=	230 nm
Increments for:		
m-OH: 2x7	=	14 nm
P-OH	=	25 nm
TOTAL	=	**269 nm**

4)

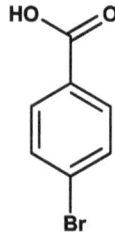

Base value for parent chromophore	=	230 nm
Br substitution at para	=	15 nm
TOTAL	=	**245 nm**

POLY-enes AND POLY-ynes

If a conjugated polyene contains more than 4 double bonds, then **Fieser-Kuhn rules** are used. According to this approach, both λ_{max} and ϵ_{max} are related to the number of conjugated double bonds as well as other structural units by the following equations.

$$\lambda_{max} = 114 \times 5M + n(48.0 - 1.7n) - 16.5 R_{endo} - 10 R_{exo}$$

$$\epsilon_{max} = (1.74 \times 10^4)n$$

n = number of conjugated double bonds

M = number of alkyl groups or substituent on the conjugated system

R_{endo} = number of rings with endocyclic double bonds in the conjugated system.

R_{exo} = number of rings with exocyclic double bonds

1) Lycopene

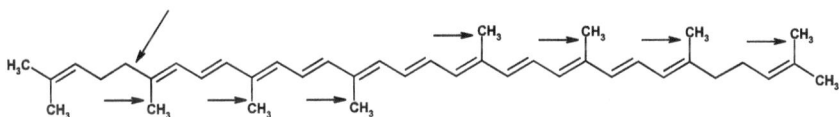

$$\lambda_{max} = 114 \times 5(8) + 11[(48.0 - 1.7(11)] - 0 - 0 = 476 \text{ nm}$$

2) β-carotene

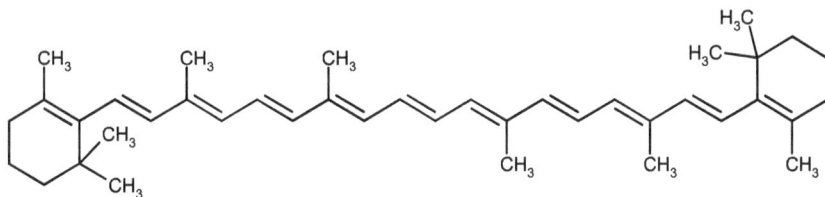

$$\lambda_{max} = 453 \text{ nm}$$

2. Exercise on UV Spectroscopy

Q 1: A natural product is known to have either structure A or B with $\lambda_{max} = 252$ nm. Which one is likely to be?

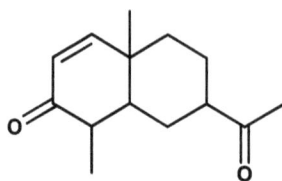 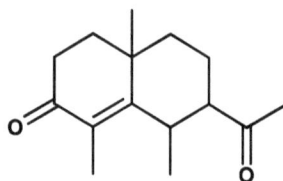

     A   B

Ans 1: **Structure A**

Base value for α, β unsaturated ketone	= 215 nm
Increments for:	
1 β- Ring residue	$1 \times 12 = 12$ nm
TOTAL	**= 227 nm**

Structure B

Base value for α, β unsaturated ketone	= 215 nm
Increments for:	
1 α-Ring residue	$1 \times 10 = 10$ nm
2 β- Ring residue	$2 \times 12 = 24$ nm
1 Exocyclic double bond	$1 \times 5\ = 5$ nm
TOTAL	**= 254 nm**

This value is closer to that observed for the natural product. Hence, the structure is B.

Q 2: Calculate λ_{max} for A and B

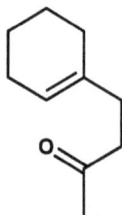 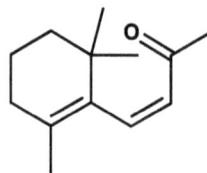

     A   B

One of the spectra has ε_{max} 20,800 cm^2/mol and the other has absorptivity max ε_{max} 10,700 cm^2/mol. How is this difference explained?

Ans 2: **Structure A**

Base value for α, β unsaturated ketone   = 215 nm

Increments for:

1 γ-Ring residue	$1 \times 18 = 18$ nm
1 γ'- Ring residue	$1 \times 18 = 18$ nm
1 Double bond extending conjugation	$1 \times 30 = 30$ nm
TOTAL	**= 281 nm**

Structure B

Base value for α, β unsaturated ketone   = 215 nm

Increments for:

1 γ-Ring residue	$1 \times 18 = 18$ nm
2 γ'- Ring residue	$2 \times 18 = 36$ nm
1 Double bond extending conjugation	$1 \times 30 = 30$ nm
TOTAL	**= 299 nm**

Structures A and B are similar except the presence of three extra methyl groups in B which cause steric hindrance in the molecule and hence decrease ε_{max}. Steric hindrance is between –CH₃ at γ' position and –H at β-position. No such hindrance is possible in A, hence, $\varepsilon_{max\ value}$ is higher in A.

Q 3: Calculate λ_{max} for the following:

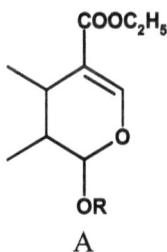

A

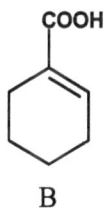

B

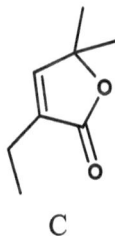

C

Ans 3:

A)

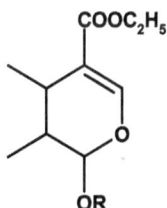

Base value for α, β unsaturated carboxylic acid and ester

= 195 nm

Increments for:

1α-Ring residue	1 × 10 = 10 nm
1β-OCH₃ residue	1 × 30 = 30 nm
TOTAL	**= 235 nm**

B)

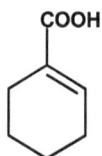

Base value for α,β unsaturated carboxylic acid and ester

= 195 nm

Increments for:

1 α-Ring residue	1 × 10 = 10 nm
1β-Ring residue	1 × 12 = 12 nm
TOTAL	**= 217 nm**

C)

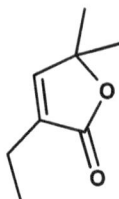

Base value for α, β unsaturated carboxylic acid and ester

= 195 nm

Increments for:

1 α-Ring residue	1 × 10 = 10 nm
1β-Ring residue	1 × 12 = 12 nm
Endocyclic α,β- double bond in 5- or 7- membered ring	= 5 nm
TOTAL	**= 222 nm**

Q 4:   Calculate λ_{max} for:

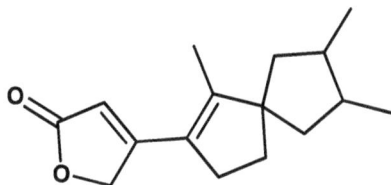

The agreement with observed value of 288 nm is poor. Explain.

Ans 4:

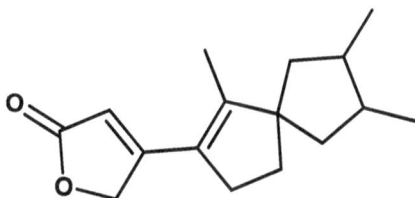

Base value for α, β unsaturated carboxylic ester

                                                                = 195 nm

Increments for:

1β-Ring residue	$1 \times 12 = 12$ nm
1 γ-Ring residue	$1 \times 18 = 18$ nm
2 γ'- Ring residue	$2 \times 18 = 36$ nm
1 Double bond extending conjugation	$1 \times 30 = 30$ nm

Endocyclic α,β- double bond in 5- or 7- membered ring

                                                                = 5 nm

TOTAL   **= 296 nm**

Observed λ_{max} is lesser than this value because of steric hindrance between α-H and γ'-CH_3 groups. Also, strain is caused due to the 5-membered ring C that causes restricted rotation due to its orientation.

Q 5: Calculate K-band absorption maxima for:

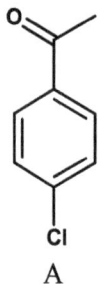

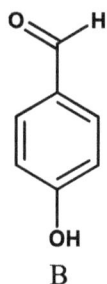

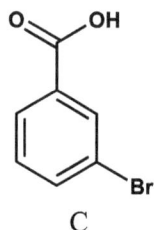

| A | B | C |

Ans 5:

A)

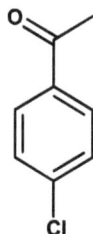

Base value for parent chromophore ArCOR

| | = | 246 nm |

Increments for:

| Cl- at para position | = | 10 nm |
| **TOTAL** | = | **256 nm** |

B)

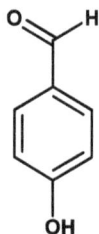

Base value for parent chromophore ArCHO

| | = | 250 nm |

Increments for:

| OH- at para position | = | 25 nm |
| **TOTAL** | = | **275 nm** |

C)

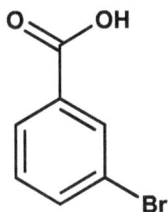

Base value for parent chromophore ArCOOH

$$= \quad 230 \text{ nm}$$

Increments for:

Br- at ortho position	=	2 nm
TOTAL	=	**232 nm**

Q 6: Calculate the λ_{max} for

a)

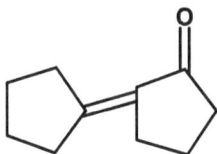

Base value for parent conjugated enone	= 215 nm
Increments for:	
1 Alkyl residue at α Position	= 10 nm
2 Alkyl residue at β Position	= 24 nm
Exocyclic double bonds to both the rings	= 10 nm
TOTAL	**= 259 nm**

b)

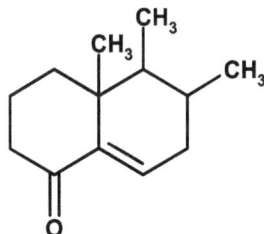

Base value for parent conjugated enone	= 215 nm
Increments for:	
1 Alkyl residue at α Position	= 10 nm
1 Alkyl residue at β Position	= 12 nm
1 Exocyclic double bond	= 5 nm
TOTAL	**= 242 nm**

c)

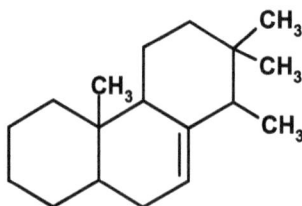

Base value for hetero annular diene	= 215 nm
Increments for:	
4 ring residues	4 × 5 = 20 nm
TOTAL	**= 235 nm**

d)

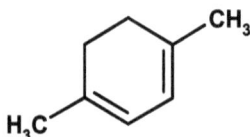

Base value for parent homo annular diene	= 253 nm
Increments for:	
4 ring residues	4 × 5 = 20 nm
TOTAL	**= 273 nm**

e)

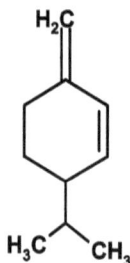

Base value for cyclic conjugated diene   = 217 nm

Increments for:

2 ring residues $2 \times 5 = 10$ nm

1 Exo cyclic double bond   = 5 nm

TOTAL   = 232 nm

f)

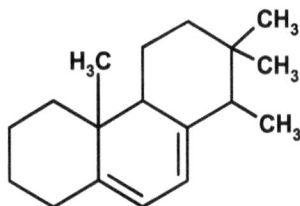

Base value for parent homo annular diene   = 253 nm

Increments for:

1 Exocyclic double bond   = 5 nm

4 ring residues $4 \times 5 = 20$ nm

TOTAL   = 278 nm

g)

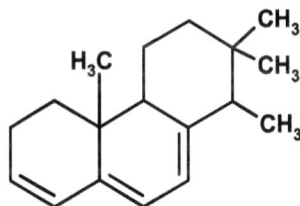

Base value for parent Homo annular diene   = 253 nm

2 Exo cyclic double bonds   = 10 nm

Double bond extending conjugation   = 30 nm

4 ring residues   = 20 nm

TOTAL   = 313 nm

h)

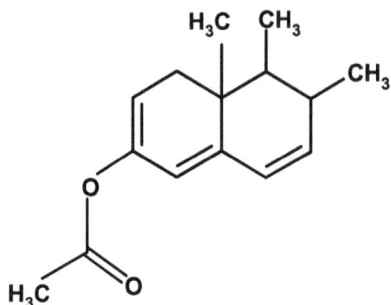

Base value for parent Homo annular diene	= 253 nm
1 Exo cyclic double bonds	= 5 nm
Double bond extending conjugation	= 30 nm
3 ring residues	= 15 nm
TOTAL	**= 303 nm**

i)

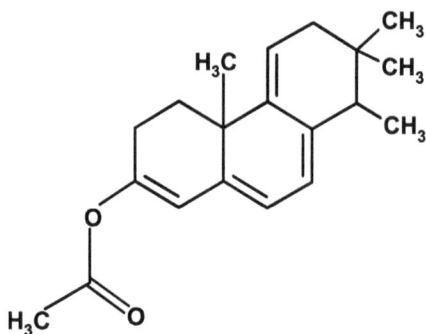

Base value for parent Homo annular diene	= 253 nm
3 Exo cyclic double bonds	= 15 nm
1-Acetyl group	= 00 nm
2 extra conjugated double bonds	= 60 nm
5 ring residues	= 25 nm
TOTAL	**= 353 nm**

2. To identify the diagnostic bands in each of the given IR spectrum and match the name from each list to proper IR Spectrum

Spectrum A-D :   1,3-Cyclohexadiene

Diphenylacetylene

1-Octene

2-Pentene

A)

Problem 3.3 Spectrum A

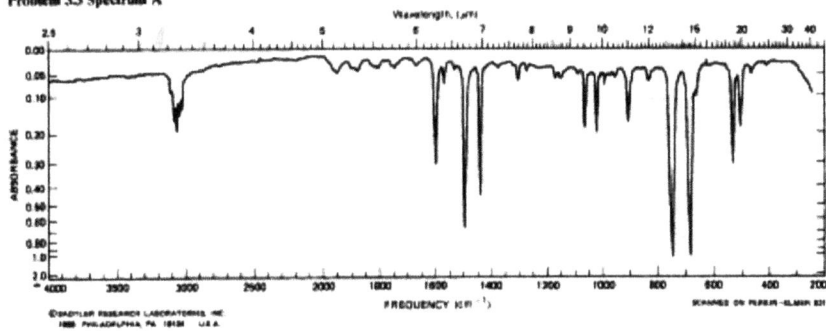

WAVENUMBER (cm^{-1})	FUNCTIONAL GROUP
3100-3020	C–H stretch
2000-1730	Overtone or combination bands
1610,1500,1450	C=C ring stretch
750	Out of plane aromatic C–H bend
690	Ring C---C bend

Because of its symmetry, no C---C band is observed in IR spectra and because of the presence of above characteristic bands, the spectra is confirmed to be of **Diphenylacetylene**.

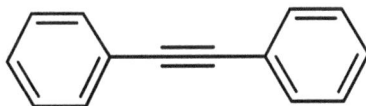

Diphenylacetylene

B)

Problem 3.3 Spectrum B

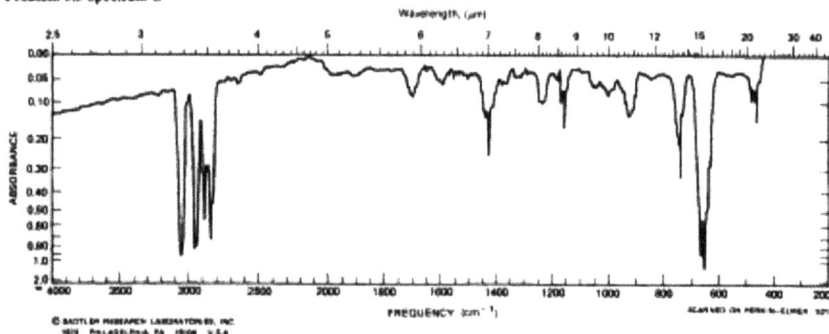

WAVENUMBER (cm⁻¹)	FUNCTIONAL GROUP
3070-2970	C–H stretch
2900	γ_{as} [asymmetrical stretch]
2850	γ_s [symmetrical stretch]
1680	Coupled C=C-C=C stretch (weak in cyclic structures)
1400	C–H in-plane bend in alkene
650	Out of plane C–H bending vibrations

Because of the presence of above characteristic bands, the spectra is confirmed to be of **1,3-Cyclohexadiene**.

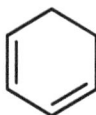

1,3-Cyclohexadiene

C)

Problem 3.3 Spectrum C

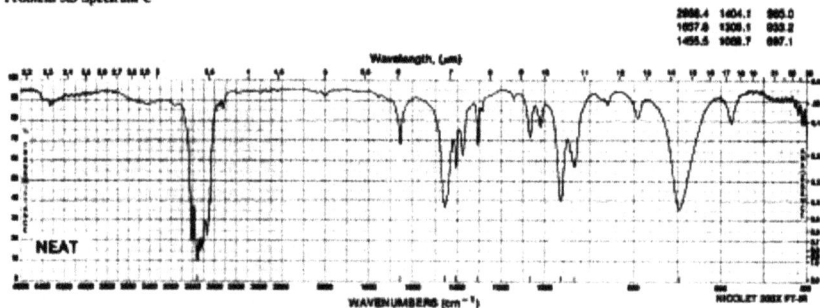

WAVENUMBER (cm⁻¹)	FUNCTIONAL GROUP
3010	C–H stretch
3000	γ_{as} CH$_3$ [asymmetrical stretch]
2960	γ_{as} CH$_2$ [asymmetrical stretch]
2920	γ_s CH$_3$ [symmetrical stretch]
2870	γ_s CH$_2$ [symmetrical stretch]
1660	C=C stretch
1450, 1400, 1370	Scissoring of CH$_2$ [δ_s], scissoring of CH$_3$ [δ_{as} ,δ_s]
960-940	Out of plane C–H bend
700	Methylene rock [ρ CH$_2$]

Because of the presence of above characteristic bands, the spectra is confirmed to be of **2-Pentene**.

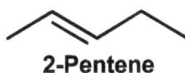

2-Pentene

D)

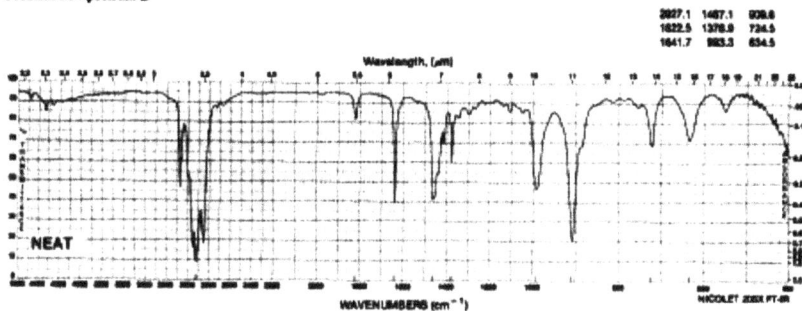

WAVENUMBER (cm⁻¹)	FUNCTIONAL GROUP
3090	Alkene C–H stretch
2950, 2910, 2850	γ_{as} CH$_3$ [asymmetrical stretch], γ_{as} CH$_2$ [asymmetrical stretch], γ_s CH$_3$ [symmetrical stretch], γ_s CH$_2$ [symmetrical stretch]
1650	C=C stretch
990, 900	Out of plane C–H bend
720	Methylene rock [ρ CH$_2$]

Because of the presence of above characteristic bands, the spectra is confirmed to be of **1-Octene.**

1-Octene

Spectra E-I :   **Butylacetate**

Butyramide

Isobutylamine

Lauric acid

Sodium propionate

E)

Problem 3.4 Spectrum E

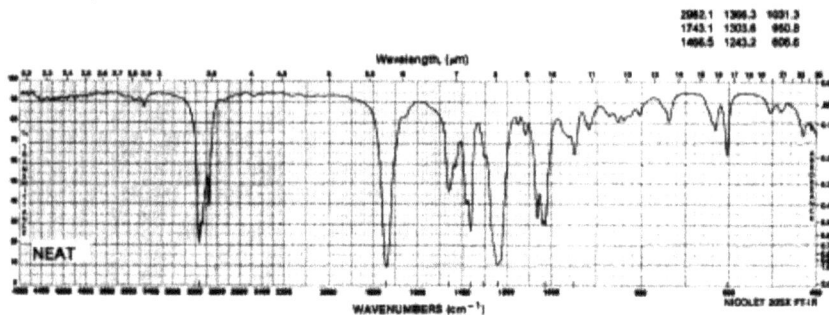

WAVENUMBER (cm^{-1})	FUNCTIONAL GROUP
2950-2870	γ_{as} CH$_3$ [asymmetrical stretch], γ_{as} CH$_2$ [asymmetrical stretch], γ_s CH$_3$ [symmetrical stretch], γ_s CH$_2$ [symmetrical stretch]
1750	-C=O stretch of ester_
1460	γ_{as} CH$_3$ [asymmetrical stretch]
1360	γ_s CH$_3$ [symmetrical stretch]
1240	Acetate C(C=O)-O stretch

Because of the presence of above characteristic bands, the spectra is confirmed to be of **Butylacetate.**

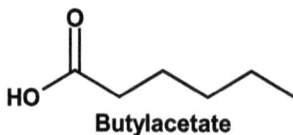

Butylacetate

F)

Problem 3.4 Spectrum F

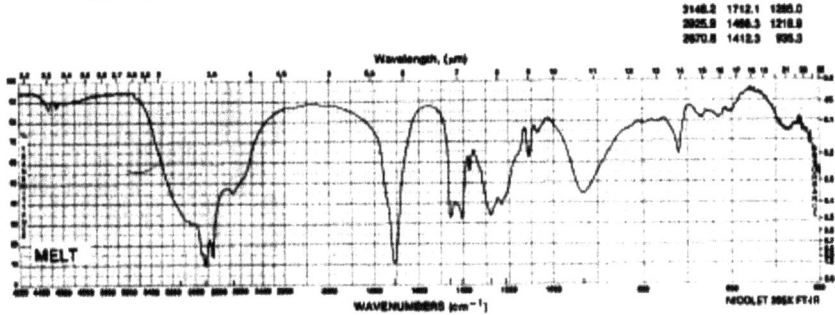

WAVENUMBER (cm⁻¹)	FUNCTIONAL GROUP
3000-2650	Broad –O-H stretch
2950, 2930, 2850	C–H stretch of CH_3, CH_2 superimposed on broad O-H stretch
1700	Normal dimeric carboxylic stretch
1460-1410	C-O-H in plane bend
1280-1230	C-O stretch of dimer
940	O-H out of plane bend

Because of the presence of above characteristic bands, the spectra is confirmed to be of **Lauric acid**.

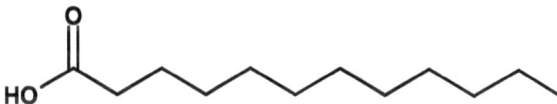

Lauric acid

G)

Problem 3.4 Spectrum G

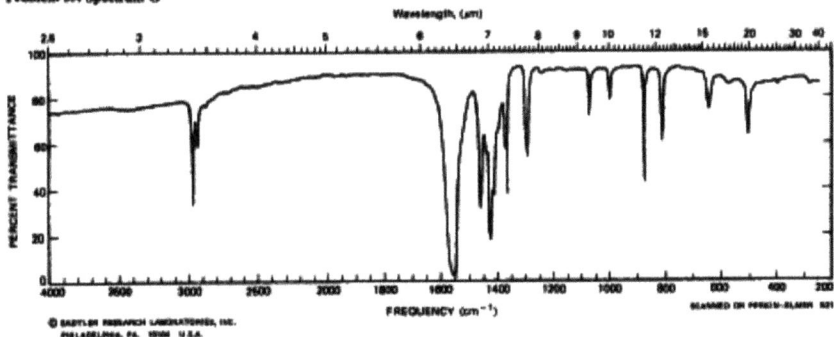

WAVENUMBER (cm^{-1})	FUNCTIONAL GROUP
2980-2920	γ_{as} CH$_3$ [asymmetrical stretch], γ_{as} CH$_2$ [asymmetrical stretch], γ_s CH$_3$ [symmetrical stretch], γ_s CH$_2$ [symmetrical stretch]
1550, 1400	C=O stretch in salts

Because of the presence of above characteristic bands, the spectra is confirmed to be of

Sodium propionate.

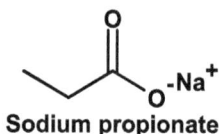

Sodium propionate

H)

Problem 3.4 Spectrum H

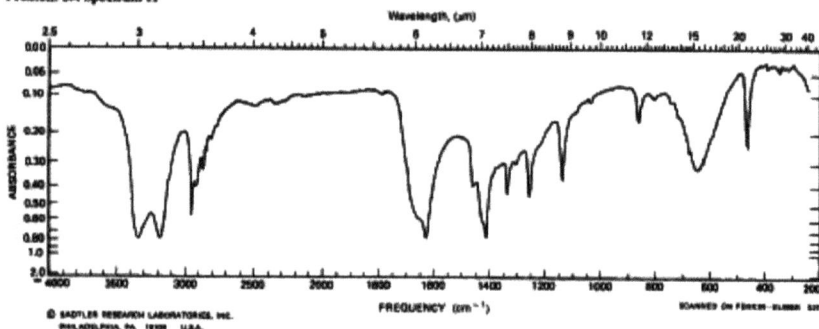

WAVENUMBER (cm^{-1})	FUNCTIONAL GROUP
3350-3170	N-H stretch coupled primary amide, hydrogen bonded- asymmetric- 3350-symmetric - 3170
2950	Aliphatic C–H stretch
1650	Overlapped C=O stretch with amide stretch band
1400	C-N stretch
650	Broad N-H out of plane bend

Because of the presence of above characteristic bands, the spectra is confirmed to be of **Butyramide.**

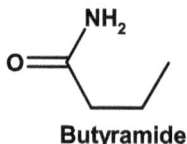

Butyramide

I)

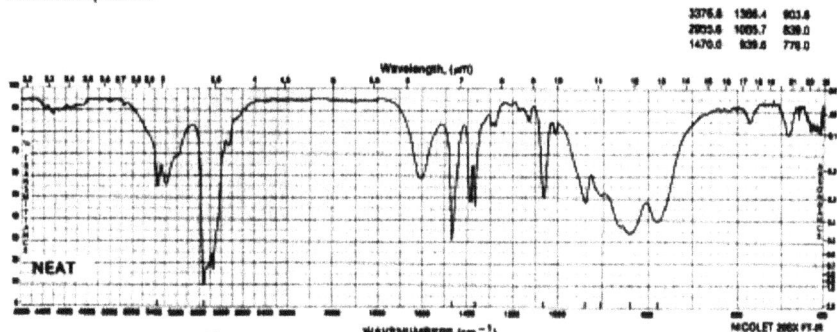

WAVENUMBER (cm^{-1})	FUNCTIONAL GROUP
3380,3300	N-H stretch, hydrogen bonded, primary amine coupled dublet: asymmetric- 3380 symmetric – 3300 [Fermi resonance band with overtone band at 1610 cm^{-1}]
2980-2880	Aliphatic C–H stretch
1610	N-H bend [scissoring]
1470	δ_s CH$_2$ [scissoring]
1060	C-N stretch
860-780	N-H wagging

Because of the presence of above characteristic bands, the spectra is confirmed to be of **Isobutylamine**.

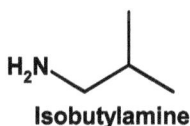

H$_2$N⤷⟨isobutyl skeleton⟩
Isobutylamine

Spectra J-M : Allyl phenyl ether

Benzaldehyde

O-Cresol

m-Toluic acid

J)

Problem 3.5 Spectrum J

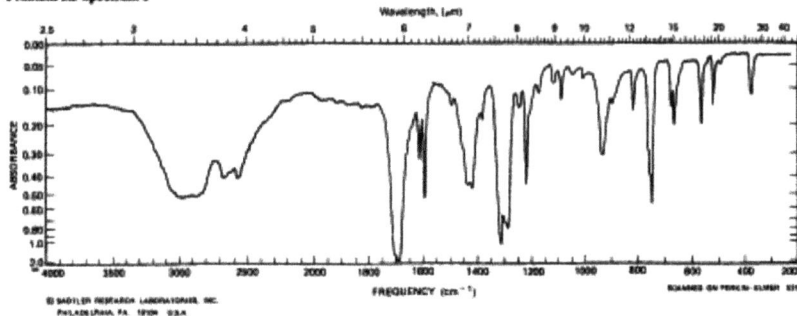

WAVENUMBER (cm⁻¹)	FUNCTIONAL GROUP
3010-2550	Broad –O-H stretch inclusive of aromatic C–H stretch (3010cm⁻¹ - 3000 cm⁻¹). This also includes C–H stretch of CH₃ superimposed upon O-H stretch
1700	Characteristic C=O stretch (carboxylic dimer)
1600	C–C ring stretch
1420	C-O-H in plane bend
1320-1280	C-O stretch of dimer
1210	Stretch, asymmetrical
930	O-H out of plane bend

Because of the presence of above characteristic bands, the spectra is confirmed to be of

m-Toluic acid.

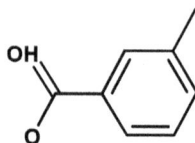

m-toluic acid

K)

Problem 3.5 Spectrum K

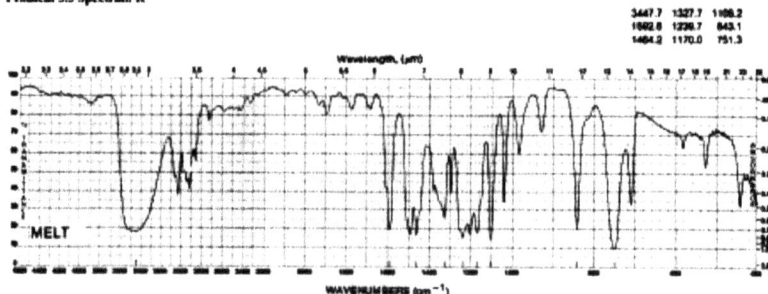

WAVENUMBER (cm⁻¹)	FUNCTIONAL GROUP
3550-3350	Broad intermolecular hydrogen bonded O-H stretch
3050	Aromatic C–H stretch
2960	C–H stretch of CH_3
1950-1680	Overtone or combination bands
1600-1460	C═C ring stretch
1330	C-O-H in plane bend
1240	C-O stretch dimer
840,750	out of plane C–H bend
710	out of plane ring C---C bend

Because of the presence of above characteristic bands, the spectra is confirmed to be of **o-Cresol**.

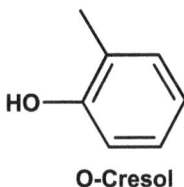

HO—

O-Cresol

L)

Problem 3.5 Spectrum L

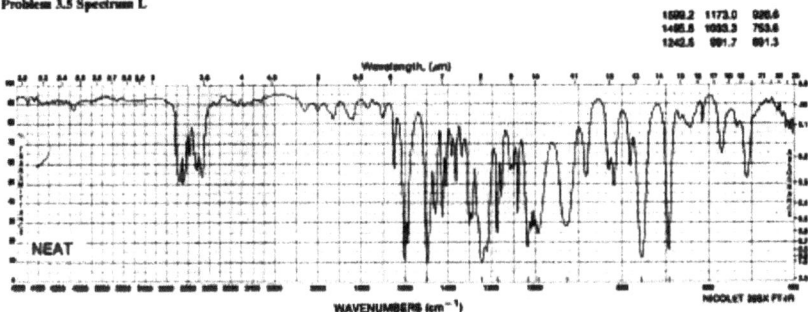

WAVENUMBER (cm⁻¹)	FUNCTIONAL GROUP
3080-3060	Aromatic C–H stretch
2920, 2880	C–H stretch of CH_3
2060-1700	Overtone or combination region
1650	Allyl group C=C stretch (monosubstituted)
1500	C═C ring stretch
1250	Asymmetric C-O-C stretch
1050	Symmetric C-O-C stretch
840,750	out of plane C–H bend
710	out of plane ring C---C bend

Because of the presence of above characteristic bands, the spectra is confirmed to be of **Allyl phenyl ether.**

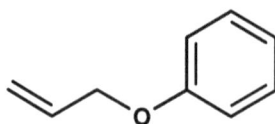

Allyl phenyl ether

M)

Problem 3.5 Spectrum M

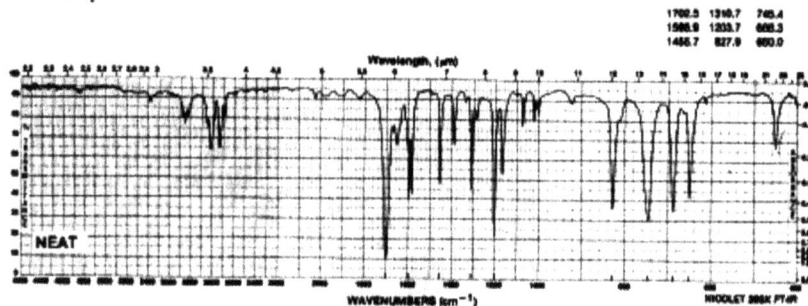

1702.5	1310.7	745.4
1586.9	1203.7	688.3
1455.7	827.9	660.0

Wavelength, (μm)

WAVENUMBERS (cm⁻¹)

NICOLET 360X FT-IR

NEAT

WAVENUMBER (cm⁻¹)	FUNCTIONAL GROUP
3070-3040	Aromatic C–H stretch
2920-2880	Aldehydic C–H stretch
2060-1800	Overtone or combination region
1700	Normal aldehydic C=O stretch
1600, 1450	C––C ring stretch
1390	Aldehydic C–H bend
750	out of plane C–H bend
690	out of plane C---C bend

Because of the presence of above characteristic bands, the spectra is confirmed to be of **Benzaldehyde**.

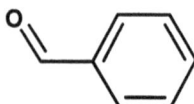

Benzaldehyde

Spectra N-R : Aniline

Azobenzene

Benzophenone oxime

Benzylamine

Dimethylamine hydrochloride

N)

Problem 3.6 Spectrum N

Merck 10,1131

3370.1	1452.3	779.2
3026.9	890.8	735.9
1604.5	858.9	698.3

Wavelength, (μm)

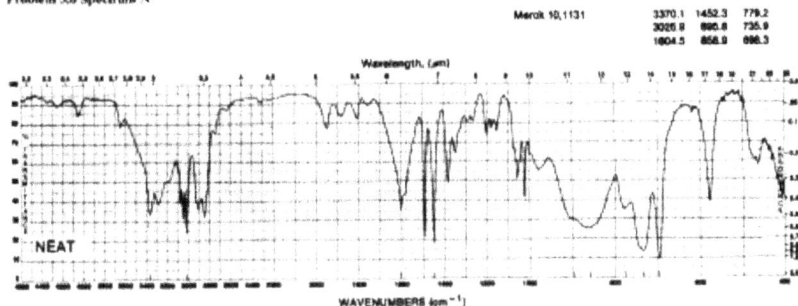

NEAT

WAVENUMBERS (cm⁻¹)

WAVENUMBER (cm⁻¹)	FUNCTIONAL GROUP
3370-3280	N-H stretch hydrogen bonded, primary amine coupled doublet: asymmetrical- 3370 Symmetrical- 3200
3060, 3030, 3000	Aromatic C–H stretch
2920, 2850	Aliphatic C–H stretch
1960-1750	Overtone or combination region
1600, 1500	C=C ring stretch
1460	CH₂ scissoring
1060-1020	C-N stretch
910-840	N-H wagging
700	out of plane C---C bend

Because of the presence of above characteristic bands, the spectra is confirmed to be of **Benzylamine**.

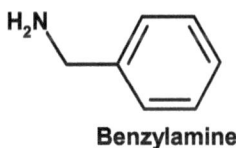

Benzylamine

O)

Problem 3.6 Spectrum O

Wavelength, (μm)

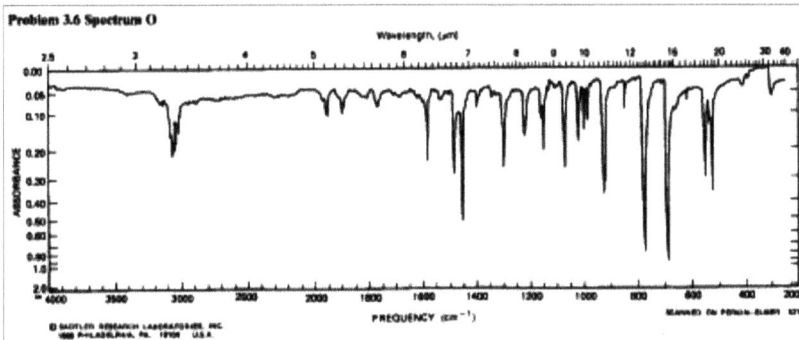

FREQUENCY (cm⁻¹)

WAVENUMBER (cm⁻¹)	FUNCTIONAL GROUP
3080	Aromatic C–H stretch
1500, 1450	C=C ring stretch
1300	O-H in plane bend
1225	O-H bend
760	out of plane aromatic C–H bend
697	out of plane ring C---C bend

Because of the presence of above characteristic bands, the spectra is confirmed to be of **Benzophenone oxime.**

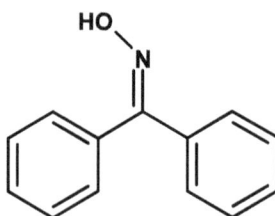

Benzopenone oxime

P)

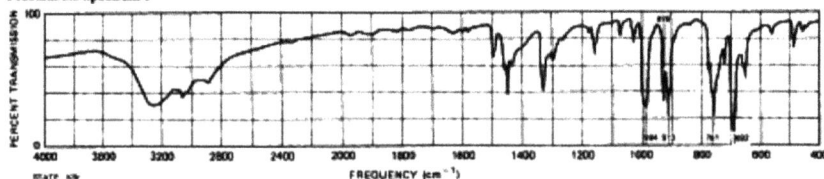

Problem 3.6 Spectrum P

WAVENUMBER (cm⁻¹)	FUNCTIONAL GROUP
3350-3100	N-H stretch of secondary amine. Shoulder on low frequency side of N-H stretch
1500-1450	N-H bend [scissoring]
1175-1020	Weak absorption band for unconjugated C-N linkage in amine
913, 692	Medium absorption N-H wagging

Because of the presence of above characteristic bands, the spectra is confirmed to be of **Dimethylamine hydrochloride.**

Dimethylamine hydrochloride

Q)

Problem 3.6 Spectrum Q

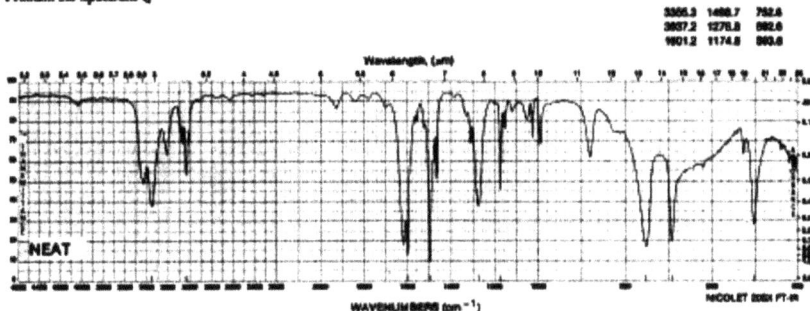

WAVENUMBER (cm⁻¹)	FUNCTIONAL GROUP
3450-3350	N-H aromatic stretch of primary amine
3200, 3050	Aromatic C–H stretch
1625-1600	N-H scissoring of primary amine
1600, 1500	C = C ring stretch
1290	Aromatic C-N stretch
750	out of plane aromatic C–H bend
690	out of plane aromatic C---C bend

Because of the presence of above characteristic bands, the spectra is confirmed to be of **Aniline**.

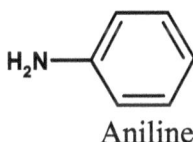

Aniline

R)

Problem 3.6 Spectrum R

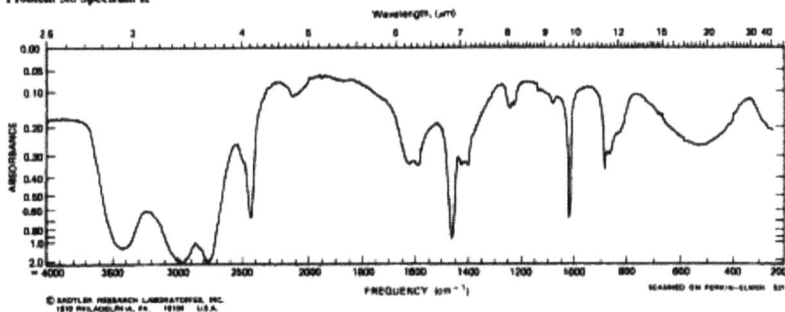

WAVENUMBER (cm^{-1})	FUNCTIONAL GROUP
3050-3000	Aromatic C–H stretch
1600	C≕C ring stretch
1575	N=N stretch (raman region)
1250-1000	Aromatic C-N stretch
650-550	Broad N-H out of plane bend

Because of the presence of above characteristic bands, the spectra is confirmed to be of **Azobenzene**.

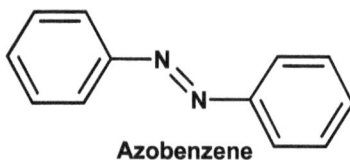

Azobenzene

3. **To identify the peaks in each of the given NMR spectra and deduce the compound**

A) $C_6H_{12}O$ (in CDCl$_3$), ketone

Peaks	Height(in cm)	No. of corresponding protons
A	0.5	$0.5 \times 4 = 2$
B	0.7	$0.7 \times 4 = 3$
C	0.5	$0.5 \times 4 = 2$
D	0.5	$0.5 \times 4 = 2$
E	0.7	$0.7 \times 4 = 3$
	TOTAL = 2.9 cm	TOTAL = 12

2.9 divisions correspond to 12 protons.

1 division correspond to $12/2.9 \approx 4$ protons

Peaks	δvalue
A	2.45 (t, 2 protons, -CH$_2$)
B	2.15 (s, no protons, -C-)
C	1.55 (q, 4 protons, 2 x –CH$_2$)
D	1.3 (s, 5 protons, -CH$_2$, -CH$_3$)
E	0.9 (t, 2 protons, -CH$_2$)

Proposed structure is

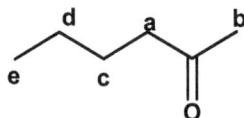

B) $C_8H_{10}O_2$, alcohol ether

Peaks	Height	No. of corresponding protons
A	0.7	0.7 x 3 = 2
B	0.7	0.7 x 3 = 2
C	0.7	0.7 x 3 = 2
D	1.0	1.0 x 3 = 3
E	0.3	0.3 x 3 = 1
	TOTAL= 3.4 cm	TOTAL = 10 protons

3.4 cm correspond to 10 protons.

1 cm correspond to $10/3.4 \approx 3$ protons

Peaks	δvalue
A	7.2 (m, aromatic)
B	6.85 (m, aromatic)
C	4.5 (s, no protons, -C-)
D	3.75 (s, no protons, -C-)
E	2.6 (s, no protons, -C-)

Proposed structure is

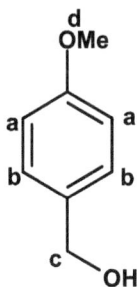

C) $C_5H_8O_2$, ester

Peaks	Height	No. of corresponding protons
A	0.5	$0.5 \times 2 = 1$
B	0.5	$0.5 \times 2 = 1$
C	0.5	$0.5 \times 2 = 1$
D	1.0	$1.0 \times 2 = 2$
E	1.4	$1.4 \times 2 = 3$
	TOTAL= 3.9 cm	TOTAL = 8 protons

3.9 divisions correspond to 8 protons.

1 division corresponds to $8/3.9 \approx 2$ protons

Peaks	δvalue
A	6.4
B	6.15 CH_2=CH-
C	5.8
D	4.2 (q, 3 protons, $-CH_3$)
E	1.3 (t, 2 protons, $-CH_2-$)

Proposed structure is

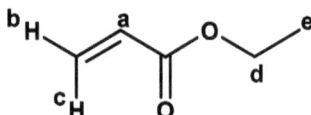

H_a and H_b have different δ values due to restricted rotation about C=C.

A) C_4H_8O

Peaks	Height (in cm)	No. of corresponding protons
A	0.4	$0.4 \times 2.4 = 1$
B	0.4	$0.4 \times 2.4 = 1$
C	2.5	$2.5 \times 2.4 = 6$
	TOTAL = 3.3 cm	TOTAL = 8 Protons

3.3 division corresponds to 8 protons.

1 division corresponds to $8/3.3 = 2.4$ protons

Peaks	δ value
A	9.65 (d, 1 proton, -CH-) due to aldehyde
B	2.45 (sept of doublets, 7 protons, 2 × –CH₃, -CH)
C	1.15 (d, 1 proton, -CH-)

Proposed structure is

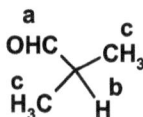

B) $C_7H_7NO_2$

Peaks	Height (in cm)	No. of corresponding protons
A	1.0	$1 \times 2 = 2$
B	1.0	$1 \times 2 = 2$
C	1.5	$1.5 \times 2 = 3$
	TOTAL = 3.5cm	TOTAL = 7 protons

3.5 cm corresponds to 7 protons

1 cm correspond to $7/3.5 = 2$ protons

Peaks	δ value
A	8.1 (m, aromatic)
B	7.3 (m, aromatic)
C	2.45 (s, no proton, -C-)

Proposed structure is

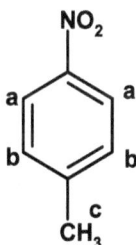

C) C_8H_9NO, amide

Peaks	Height (in cm)	No. of corresponding protons
A	0.5	0.5 × 2 = 1
B	0.9	0.9 × 2 = 2
C	0.9	0.9 × 2 = 2
D	0.4	0.4 × 2 = 1
E	1.3	1.3 × 2 = 3
	TOTAL = 4.0 cm	TOTAL = 9 protons

4 divisions correspond to 9 protons

1 division correspond to 9/4 = 2.5 protons

Peaks	δ value
A	9.9 (s, no protons, -C-)
B	7.6 (m)
C	7.3 (m)
D	7.0 (m)
E	2.05 (s, no protons, -C-)

Proposed structure is

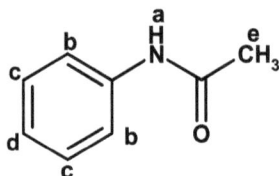

D) H_8O, ether

Peaks	Height (in cm)	No. of corresponding protons
A	0.4	0.4 × 2.5 = 1
B	0.4	0.4 × 2.5 = 1
C	0.4	0.4 × 2.5 = 1
D	0.9	0.9 × 2.5 = 2
E	1.1	1.1 × 2.5 = 3
	TOTAL = 3.2 cm	TOTAL = 8 protons

3.2 divisions corresponds to 8 protons

1 division corresponds to 8/3.2 = 2.5

Peaks	δ value
A	6.45 (d of d, 2 protons, 2 × –CH)
B	4.15 (d of d, 2 protons, 2 × –CH)
C	4.00 (d of d, 2 protons, 2 × –CH)
D	3.75 (q, 3 protons, -CH₃)
E	1.30 (t, 2 protons, -CH₂-)

Proposed structure is

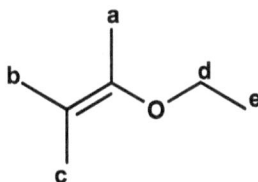

E) $C_{10}H_{12}O_2$, ester

Peaks	Height (in cm)	No. of corresponding protons
A	0.6	0.6 × 3.25 = 2
B	0.3	0.3 × 3.25 = 1
C	0.6	0.6 × 3.25 = 2
D	0.6	0.6 × 3.25 = 2
E	0.6	0.6 × 3.25 = 2
F	1.0	1.0 × 3.25 = 3
	TOTAL = 3.7 divisions	TOTAL = 12 protons

3.7 divisions correspond to 12 protons

1 division correspond to 12/3.7=3.25 protons

Peaks	δ value
A	8.05 (m, aromatic)
B	7.55 (m, aromatic)
C	7.4 (m, aromatic)
D	4.3 (t, 2 protons, -CH₂-)
E	1.8 (sextet, 5 protons, -CH₃, -CH₂)
F	1.0 (t, 2 protons, -CH₂-)

Proposed structure is

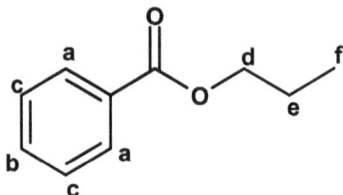

5. To find δ value of the following ^{13}C compounds

C^{13} Atom	Shift (ppm)
α	+9.1
B	+9.4
γ	−2.5
Δ	+0.3
ε	+0.1
$1°(3°)$	−1.1
$1°(4°)$	−3.4
$2°(3°)$	−2.5
$2°(4°)$	−7.2
$3°(2°)$	−3.7
$3°(3°)$	−9.5
$4°(1°)$	-1.5
$4°(2°)$	-8.4

1)

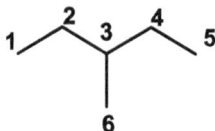

$δ = -2.5 + ΣnA$

Methane absorbs at -2.5 ppm.

n → No. of carbon atoms

A → Additive shift parameter

- For carbon atom 1 : We have 1α, 1β, 2γ, 1δ carbons.

$\delta_1 = -2.5 + (9.1 \times 1) + (9.4 \times 1) + (-2.5 \times 2) + (0.3 \times 1) = 11.3$

- For carbon atom 2 : We have 2α, 2β, 1γ and it is a 2° carbon with a 3° carbon attached to it.

$[2°(3°) = -2.5]$

$\delta_2 = -2.5 + (9.1 \times 2) + (9.4 \times 2) + (-2.5 \times 1) + (-2.5 \times 1) = 29.5$

- For carbon atom 3 : We have 3α, 2β, carbons and it is a 3° carbon with two 2° carbons attached $[3°(2°) = -3.7]$ to it.

$\delta_3 = -2.5 + (9.1 \times 3) + (9.4 \times 2) + (-3.7 \times 2) = 36.2$

- For carbon atom 6 : We have 1α, 2β, 2γ carbons and it is a 1° carbon with a 3° carbon attached : $[1°(3°) = -1.1]$

$\delta_6 = -2.5 + (9.1 \times 1) + (9.4 \times 2) + (-2.5 \times 2) + (-1.1 \times 1) = 19.3$

2) To find out the effect of substituent (OH) on δ values.

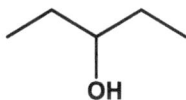

OH

	C 1	C 2	C 3	
Pentane	13.9	22.8	34.7	

	α		β		γ
	Terminal	Internal	Terminal	Internal	
OH	+48	+41	+10	+8	–5

- For α carbon: As it is carbon number 3 from left, hence, base value = 34.7 for pentane molecule.

As –OH is attached at α-position internally, hence, effect of substituent adds 41 ppm.

$\delta_{C\alpha} = 34.7 + 41 = 75.8$ ppm

- For β carbon: As it is carbon number 2 from left, hence, base value = 22.8 for pentane molecule.

As –OH is attached at β-position internally, hence, effect of substituent adds 8 ppm.

$\delta_{C\beta} = 22.8 + 8 = 30.8$ ppm

- For γ carbon: As it is carbon number 1 from left, hence, base value = 13.9 for pentane molecule.

As -OH is attached at γ-position internally, hence, effect of substituent adds –5 ppm.

$\delta_{C\gamma} = 13.9 - 5 = 8.9$ ppm

3)

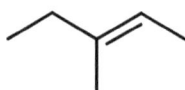

α	+ 10.6
β	+ 7.2
γ	– 1.5
α′	– 7.9
β′	– 1.8
γ′	– 1.5
z(cis)correction	– 1.1

$\delta_{C-3} = 123.3 + (2 \times 10.6) + (1 \times 7.2) + (1 \times - 7.9) \times 1.1 = 142.7$ ppm

Here, $123.3 \rightarrow$ Shift for ethylene group

$(2 \times 10.6) \rightarrow$ as C-3 has 2 α-carbons each having value = + 10.6.

$(1 \times 7.2) \rightarrow$ as C-3 has 1 β-carbon having value = + 7.2

$(1 \times - 7.9) \rightarrow$ as there is 1 α′ carbon having value = – 7.9

$- 1.1 \rightarrow$ z (cis) correction

Similarly,

$\delta_{C-2} = 123.3 + (1 \times 10.6) + (2 \times -7.9) + (1 \times 1.8) - 1.1 = 115.2$ ppm

4. To indentify the compounds by the interpretation of different spectra

A) Compound 1

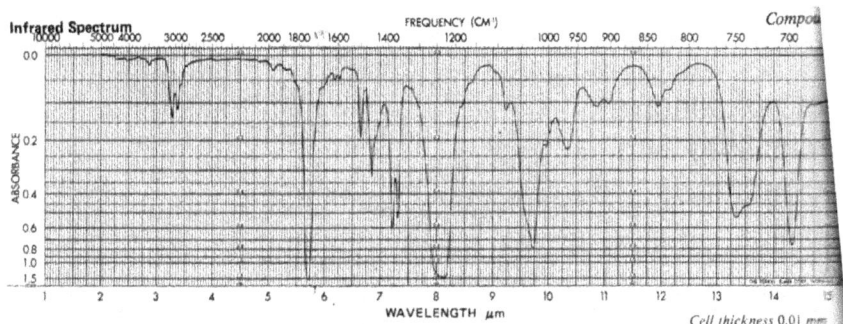

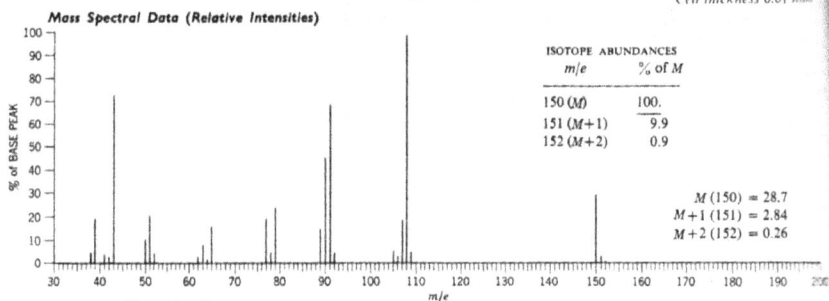

Ultraviolet Data

λ_{max}^{EtOH}	ϵ_{max}		
		252	153
268	101	248 (s)	109
264	158	243 (s)	78
262	147		
257	194	(s) = shoulder	

ISOTOPE ABUNDANCES

m/e	% of M
150 (M)	100.
151 (M+1)	9.9
152 (M+2)	0.9

M (150) = 28.7
M+1 (151) = 2.84
M+2 (152) = 0.26

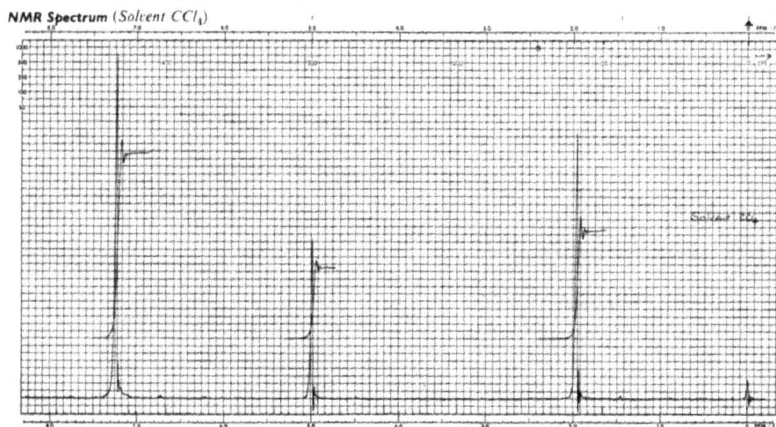

seven sets of spectra translated into compounds

- Assign molecular formula

m/z	% of M	M = 150
150 (M)	100	M+1 = 9.9
152 (M+2)	0.9	M+2 = 0.9

Short listing from the appendix

M=150	M+1	M+2
$C_7H_{10}O_4$	9.25	0.38
$C_8H_8NO_2$		
$C_8H_{10}N_2O$	9.61	0.61
$C_8H_{12}N_3$		
$C_9H_{10}O_2$	9.96	0.84
$C_9H_{12}NO$	10.71	0.52
$C_9H_{14}N_2$		

The molecular formula which can be assigned is $C_9H_{10}O_2$.

- From IR Spectra

Wavenumber (cm^{-1})	Functional group
1745	C=O stretch
1600-1400	C-O-C stretch
1225	C=C out of plane bend which shows that it is monosubstituted
749, 697	C—H stretch

- From UV Spectra

From the given UV data, it can be seen that benzene is not in conjugation with C –O.

- From NMR Spectra

Peaks	Height (in cm)	No. of corresponding protons	δ value
A	5.8	5.8 X 0.83 = 4.8 ≈ 5	7.22 (m, 5 protons, aromatic)
B	2	2 X 0.83 = 1.6 ≈ 2	5 (s, 2 protons)
C	4.2	9.2 X 0.83 = 3.4 ≈ 3	2 (s, 3 protons, -CH$_3$)
	TOTAL=12 cm	TOTAL = 10 protons	

Result: The structure for compound is

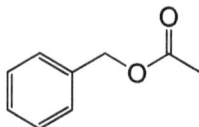

Benzyl acetate

A) Compound 2

Infrared Spectrum

FREQUENCY (CM⁻¹)

10000   5000 4000   3000 2500   2000 1800 1600   1400   1200   1000 950 900   850   800   750   70C

ABSORBANCE

0.0
0.2
0.4
0.6
0.8
1.0
1.5

1   2   3   4   5   6   7   8   9   10   11   12   13   14   15

Mass Spectral Data (Relative Intensities)   WAVELENGTH μm

Cell thick-ness

% of BASE PEAK

100
90
80
70
60
50
40
30
20
10
0

30   40   50   60   70   80   90   100   110   120   130   140   150   190   20C

m/c

ISOTOPE ABUNDANCES

m/e	% of M
126 (M)	100.
127 (M+1)	7.02
128 (M+2)	0.81

$MW = 126.033$
$M(126) = 31.95$
$M+1(127) = 2.24$
$M+2(128) = 0.26$

Ultraviolet Data

λ_{max}^{EtOH}	$\log \epsilon_{max}$
220.0 (s)	3.47
250.5	4.13

(s) = shoulder

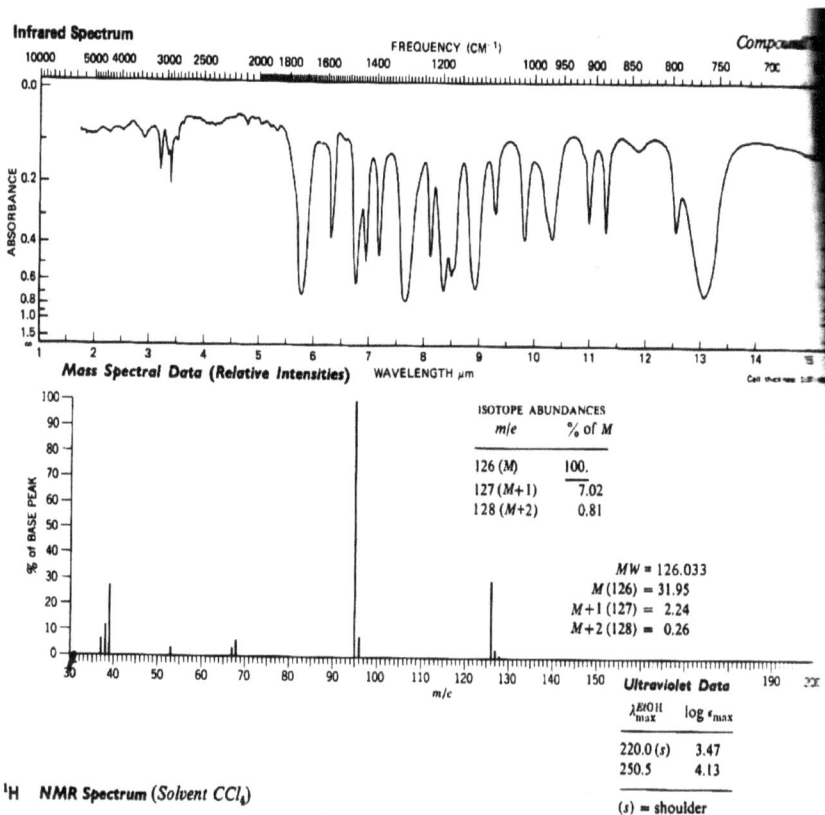

¹H   NMR Spectrum (Solvent CCl₄)

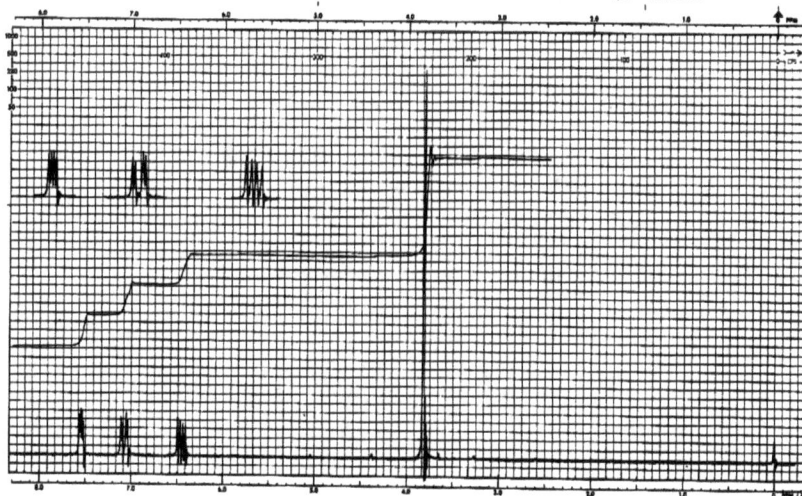

- Assign molecular formula

  Isotope abundance

m/z	% of M
126 (M)	100
128 (M+2)	0.81

  M = 121
  M + 2 = 0.81

- From IR Spectra

Wavenumber (cm^{-1})	Functional group
3150	C–H stretching
1730	Intense stretching band of ring
1600-1400	C=O stretch band

- From UV data

  The intense band in UV Spectra at 250.5nm suggests that C=O is in conjugation with aromatic ring.

- From NMR Spectra

Peaks	Height (in cm)	No. of corresponding protons	δ value
A	3.4	1.01 × 3.4 = 3.43 ≈ 3	3.95 (t, 3 protons)
B	0.6	0.6 × 3.4 = 0.7 ≈ 1	6.5 (s, 1 proton)
C	0.5	0.5 × 3.4 = 0.6 ≈ 1	7.2 (s, 1 proton)
D	0.6	0.6 × 3.4 = 0.7 ≈ 1	7.6 (s, 1 proton)
	TOTAL = 5.1cm	TOTAL = 6 protons	

- From C^{13} NMR Spectra

  Positive peaks are due to –CH or CH$_3$. If there is DEPT 90, then peaks are due to CH; but since there is none in DEPT 90, hence, the peak is due to CH$_3$.

  Result: The structure for the compound is

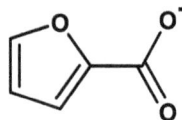

  Methyl-2-furoate

B) Compound 3 [Molecular Formula: $C_9H_{11}NO$]

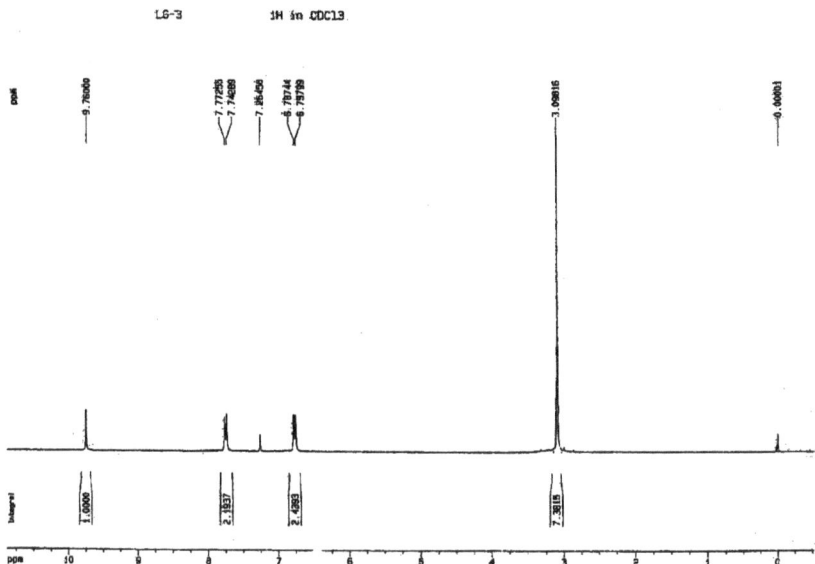

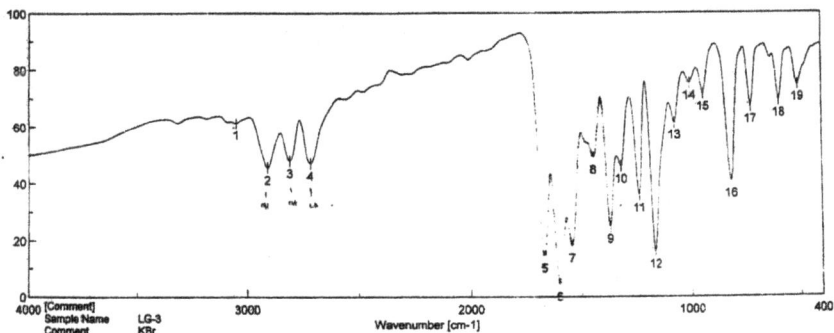

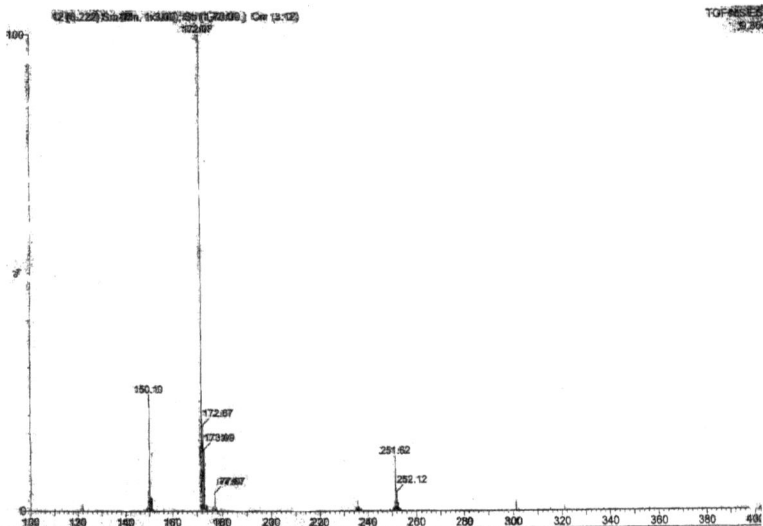

Spectral Analysis:

IR (KBr, cm^{-1}): 1668, 3049, 2928 and 2809

H^{1}NMR (CDCl$_3$, 300 MHz): δ = 3.09816 (s, 6H),

6.7579-6.7874 (d, 2H, J = 9 Hz)

7.7428-7.7725 (d, 2H, J = 9 Hz)

9.7600 (s, 1H)

ESI MS: m/z = 149 [M$^+$]

172 [M+Na]

The peak at 1669 cm^{-1} in IR spectrum reveals the presence of carbonyl peak. Further peaks at 3049, 2928 and 2809 cm^{-1} show the presence of aliphatic and aromatic groups.

H^{1}NMR spectrum shows that there are in total 11 protons and are of 4 types (2 singlet and 2 doublet).

Singlet at δ 3.0986 ppm corresponds to 6 alkyl protons and has shifted upfield, may be due to presence of hetero atom (N) attached to it (probably dimethyl amino group).

2 doublets at δ 6.75-6.78 ppm (ortho to carbonyl group) and δ 7.74-7.77 ppm (ortho to dimethyl amino group) with a coupling constant of 9.0 indicate they are aromatic protons adjacent to each other.

The singlet at δ 9.76 ppm corresponds to aldehydic proton.

Mass spectrum shows the molecular ion peak at M/z 149 corresponding to dimethyl amino benzaldehyde. It also show peak at m/z 172 which corresponds to M+Na.

Para or meta or ortho dimethyl amino benzaldehyde are possible.

It is confirmed that is p-dimethyl amino benzaldehyde (I) as it has given only doublets but not triplets which is only possible in ortho and meta derivatives (II & III).

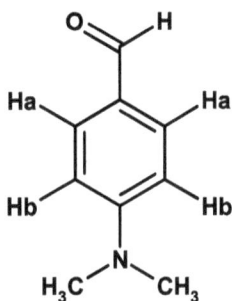

I

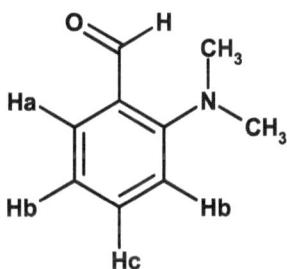

II

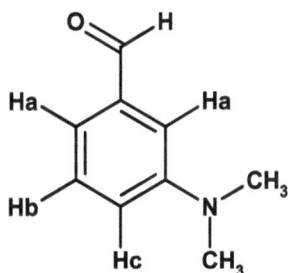

III

C) Compound 4 Molecular Formula: $C_{10}H_{12}O_4$

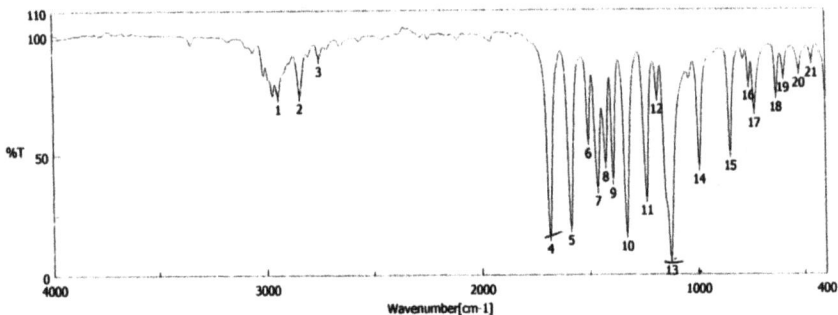

Wavenumber[cm-1]

Accumulation   45
Zero Filling   ON
Gain   4
Date/Time   11/19/07 4:58PM
Update   11/19/07 5:02PM
Operator   SATYA
File Name   Memory95
Sample Name   LG_4
Comment   KBr

Resolution   4 cm-1
Apodization   Cosine
Scanning Speed   2 mm/sec

1: 2944.77, 74.3322   2: 2843.52, 74.4924   3: 2752.89, 60.0347   4: 1685.48, 16.2942
5: 1587.13, 19.8966   6: 1504.2, 56.5108   7: 1491.78, 35.9903   8: 1425.14, 45.8082
9: 1391.39, 38.2343   10: 1329.68, 17.3424   11: 1236.16, 32.172   12: 1166.97, 73.9879
13: 1127.19, 7.10368   14: 891.232, 44.8346   15: 845.633, 50.4564   16: 757.888, 79.7307
17: 730.889, 68.2746   18: 627.716, 74.7736   19: 583.965, 82.908   20: 521.85, 84.8619
21: 463.796, 89.2075

LG-4   1H in CDCl3   21.11.07

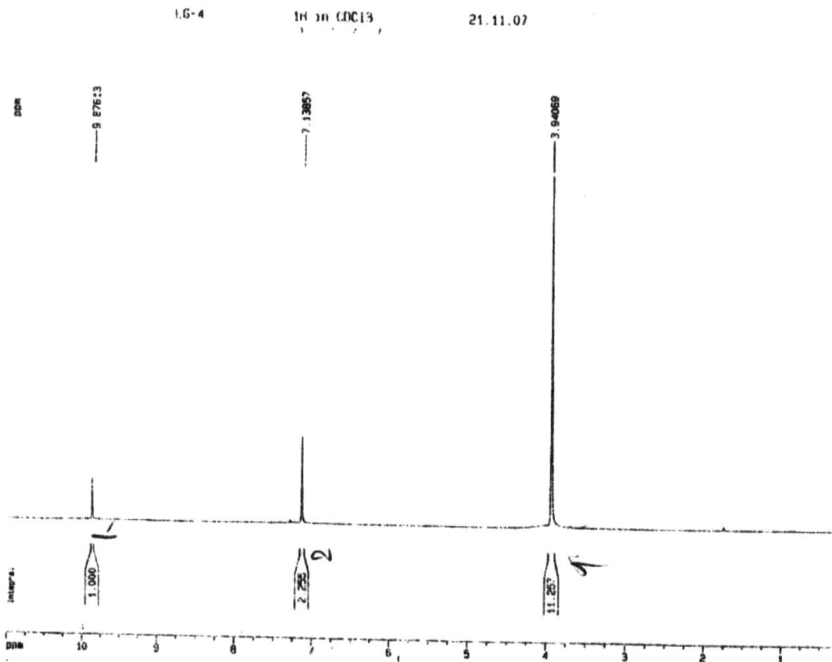

14 (0.259) Sm (Mn, 1x3.00); Sb (1,70.00); Cm (3:14)

100

219.08

T₁

%

251.11

220.08

169.10

197.10

252.11

159.10

301.17

322.10

479.

0

| 100 | 120 | 140 | 160 | 180 | 200 | 220 | 240 | 260 | 280 | 300 | 320 | 340 | 360 | 380 | 400 | 420 | 440 | 460 | 48(|

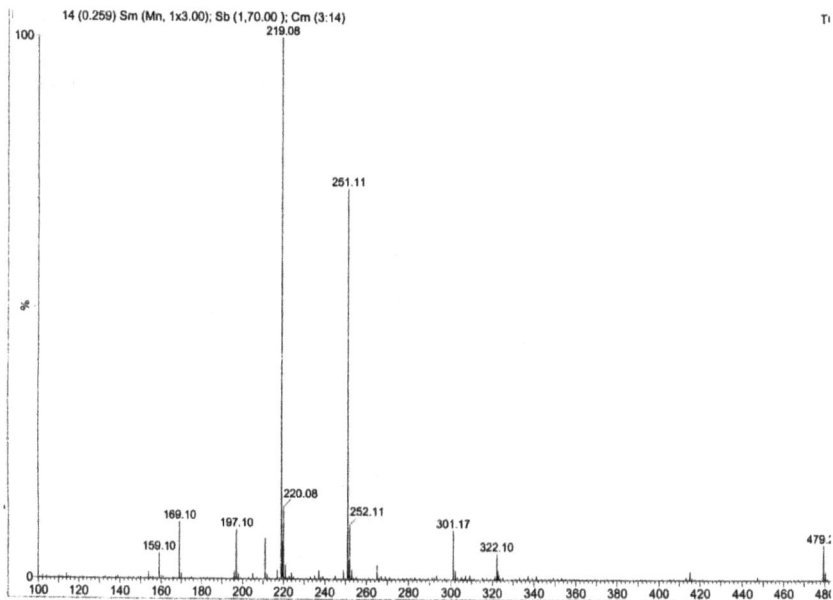

Spectral Analysis:

IR (KBr, cm⁻¹): 1685 (carbonyl Group), 2944, 2843, 2752 (Aromatic and aliphatic groups)

H¹NMR (CDCl₃, 300 MHz): δ = 3.940 (s, 9H), 7.138 (s, 2H), 9.876 (s, 1H).

ESI-MS: m/z = 196 [M⁺], 219 [M+Na]

The peak at 1685 cm⁻¹ reveals the presence of aldehydic carbonyl peak. Further 2944, 2843 and 2752 peaks indicate aromatic and aliphatic groups.

The H¹NMR spectrum shows that there are totally 12 protons of 3 types viz aliphatic protons [nine protons] (singlet at δ 3.94 ppm) shifted to upfield due to presence of hetero atom adjacent to it (O).

The singlet at δ 7.13 ppm indicates the aromatic protons (2 protons) ortho to aldehydic and methoxy groups.

The singlet at δ 9.87 corresponds to aldehydic proton.

The mass spectrum shows the molecular ion peak corresponding to its molecular weight at m/z 196 and also records a peak at m/z 219 corresponding to (M+Na).

Possible structures are 3,4,5-trimethoxy benzaldehyde (1), 2,4,6-trimethoxy benzaldehyde (2), 2,3,4-trimethoxy benzaldehyde (3), and 2,4,5-trimethoxy benzaldehyde (4).

It has been confirmed that the given molecule is Structure 1 based on the following facts

In structure 2 the methoxy groups (2 methoxy groups at ortho position and one methoxy group at para position) present in the molecule are not equivalent and thus it should give 2 different peaks for methoxy groups at ortho and para positions respectively. However the spectra shows equivalent protons thus structure 2 can be ruled out.

In structure 3, the methoxy groups are not equivalent the para methoxy group should give doublet and methoxy at ortho & meta positions should give singlets. Thus structure 3 is rejected.

Similarly structure 4 is rejected based on the fact that they must appear as 3 different peaks as methoxy groups are not equivalent as they are in different environment.

In structure 1, all the methoxy groups are equivalent and appear as singlet at δ 3.94 ppm and 2 aromatic protons ortho position appears as singlet at δ 7.13 ppm.

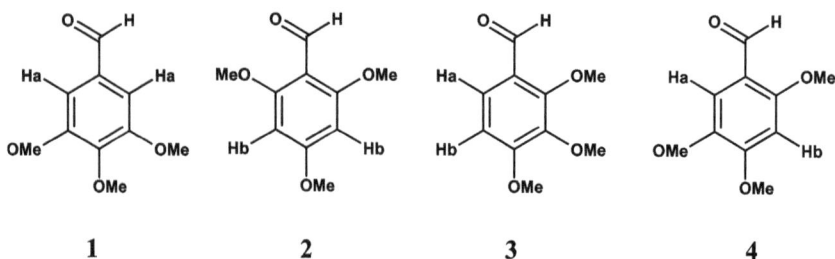

| 1 | 2 | 3 | 4 |

D) **Compound 5** Molecular Formula: $C_{16}H_{14}N_2O_2$

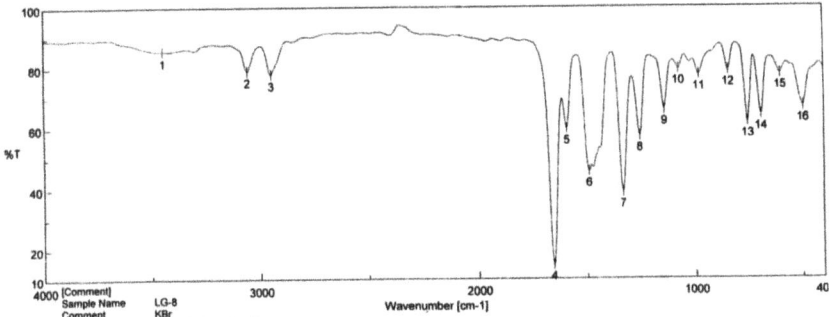

[Comment]
Sample Name   LG-8
Comment   KBr
User   SANDIP CHOWDHURY
Division   CHEMISTRY
Company   IICB

[Data Information]
Creation Date   11/21/2007 2:53 PM
Data array type Linear data array
Horizontal   Wavenumber [cm-1]
Vertical   %T
Start   349.053 cm-1
End   7800.65 cm-1
Data pitch   0.964233 cm-1
Data points   7729

Result of Peak Picking

No.	Position	Intensity	No.	Position	Intensity	No.	Position	Intensity
1	3448.1	85.4321	2	3057.58	76.8572	3	2947.66	77.7463
4	1653.66	14.7226	5	1591.95	59.6919	6	1466.78	45.6439
7	1332.57	38.6474	8	1253.5	57.0745	9	1142.62	65.9291
10	1076.08	79.1696	11	981.59	77.1972	12	843.704	78.6285
13	755.959	61.7054	14	695.212	64.0747	15	804.574	77.4663
16	497.544	86.7275						

LG-B   1H in CDCl3   20.11.07

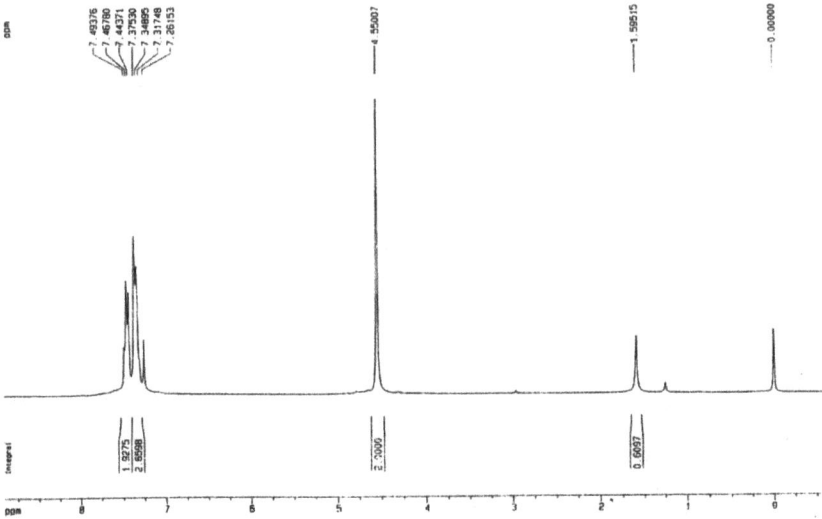

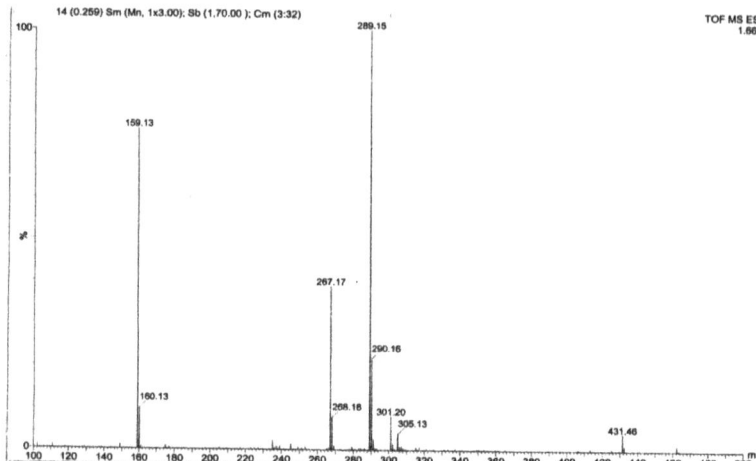

14 (0.259) Sm (Mn, 1x3.00); Sb (1.70.00); Cm (3:32)   TOF MS ES
1.66

289.15
159.13
267.17
290.16
160.13
268.16 301.20
305.13
431.46

Spectral Analysis:

IR (KBr, cm^{-1}): 1653 (Amide Carbonyl), 2947, 3057 (Aromatic groups)

H^{1}NMR (CDCl$_3$, 300 MHz): δ = 4.5507(s. 4H), 7.2615-7.4937 (m, 10H)

ESI-MS: m/z = 267 [M+1], 289 [M+Na]

The IR spectrum shows peak at 1653 cm^{-1} corresponding to amide carbonyl. Further the peaks at 2947 cm^{-1} and 3057 cm^{-1} indicates the presence of aryl groups.

The H^{1}NMR Spectrum reveals the presence of totally 14 protons of 2 varieties (aliphatic and aromatic).

The singlet at δ 4.55 ppm corresponds to 4 alkyl type protons which has got shifted to upfield (May be due to presence of electronegative functional groups like N and C=O groups)

The multiplet at δ 7.26-7.49 ppm corresponds to 10 aromatic protons which may be due to 2 phenyl rings.

Further the Mass spectra shows molecular ion peak (M+1) at m/z 267 corresponding to its molecular weight and (M+Na) peak at m/z 289.

The following 3 structures have been proposed based on the following analogy.

Structure 3 shows symmetry in one plane and more over there will be coupling to the protons from adjacent methylene protons and would have given triplet. But the spectra is showing singlet at δ 4.55 ppm. Thus this structure is ruled out.

Structure 2 also exhibits symmetry at one axis but the H_a protons; adjacent to carbonyl groups has more desheilding effect than that of H_b protons adjacent to 'N' thus should give 2 different peaks. However the NMR spectrum shows a singlet which won't fit into the above structure thus rejected.

The structure 1 is also having axis of symmetry similar to other structures but the protons are equivalent giving rise to singlet at δ 4.55 ppm which confirms the structure of the given compound.

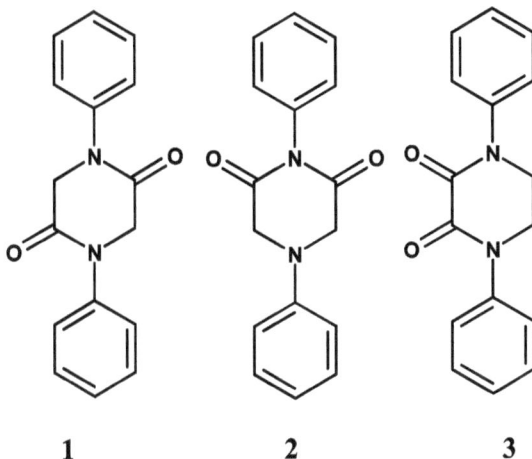

| 1 | 2 | 3 |

E) **Compound 6** Molecular Formula: $C_{15}H_{10}O_7$

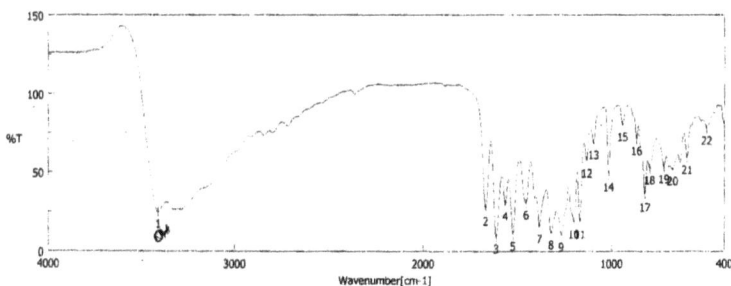

Accumulation 65
Zero Filling ON
Gain 4
Date/Time 11/20/07 5:04PM
Update 11/20/07 5:06PM
Operator SATYA
File Name Memory#5
Sample Name LG-11
Comment KBr

Resolution 4 cm-1
Apodization Cosine
Scanning Speed 2 mm/sec

1: 3406.67, 24.8067   2: 1666.2, 36.338   3: 1610.27, 8.96412   4: 1583.82, 29.1790
5: 1522.52, 10.5689   6: 1453.1, 30.1601   7: 1382.71, 16.3312   8: 1319.87, 11.3774
9: 1265.07, 10.0974   10: 1200.47, 18.0324 11: 1108.85, 15.0466 12: 1132.01, 96.7499
13: 1094.41, 66.1403 14: 1014.37, 47.8411 15: 939.183, 79.8123 16: 864.917, 70.5947
17: 822.491, 35.0235 18: 796.457, 82.3592 19: 723.176, 63.1544 20: 676.82, 51.7086
21: 602.646, 59.9476 22: 494.852, 77.4969

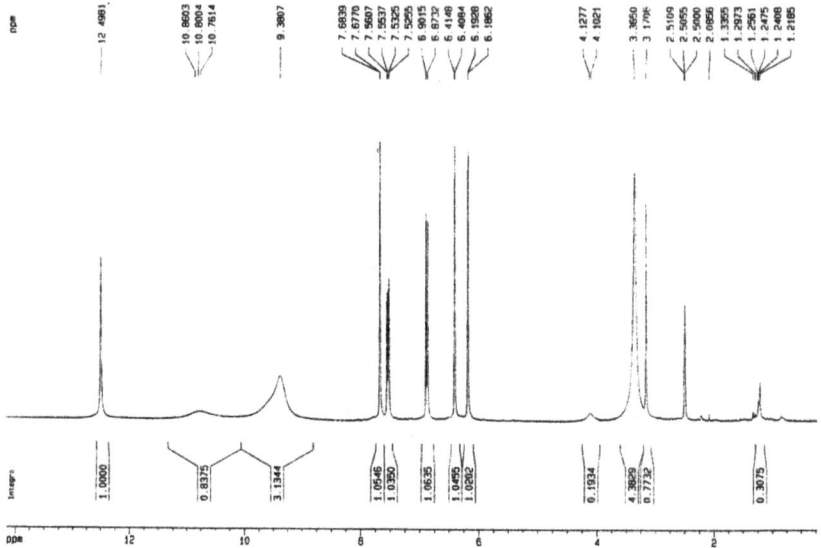

LG-1:   1H in DMSO-d6   22.11.07

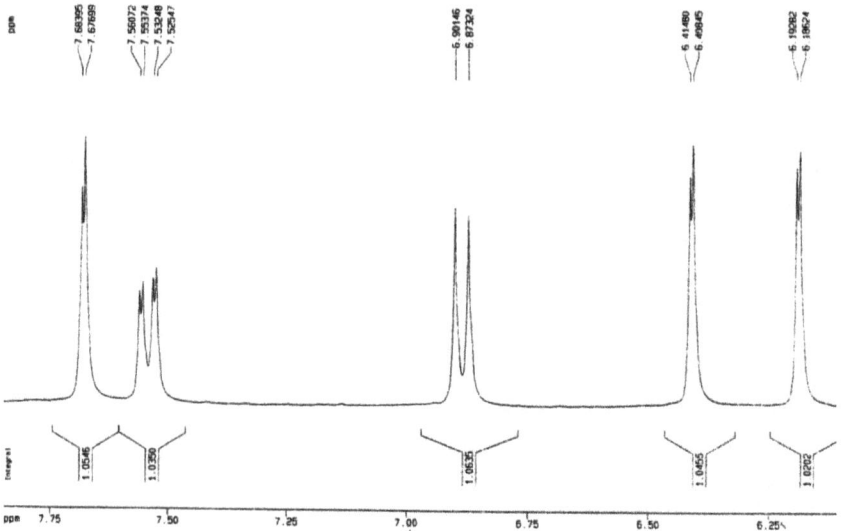

13 (0.241) Sm (Mn. 1x3.00); Sb (1,70.00); Cm (3:13)

TOF MS
1.

325.06

301.17

303.07

326.06

114.95

132.10   161.00

217.09 237.13 253.16

265.17

307.22 337.21

429.26

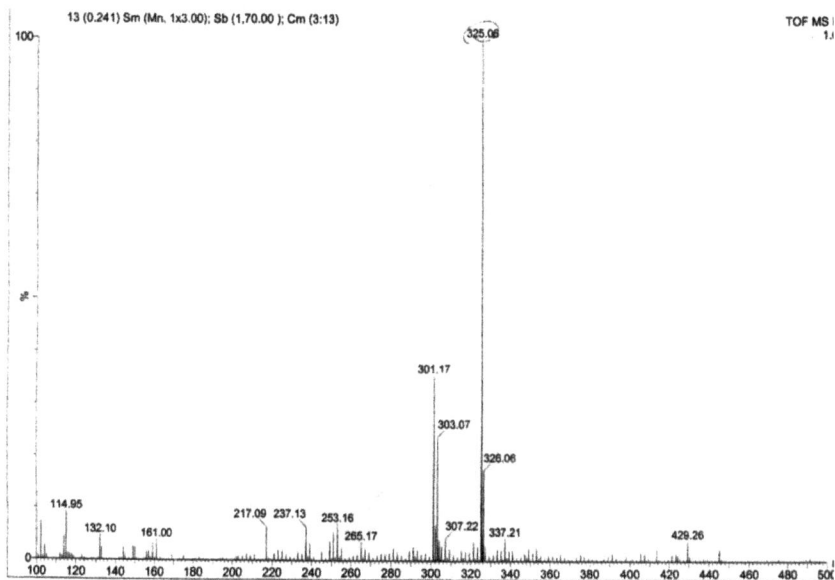

LG-11   1H in DMSO-d6   22.11.07

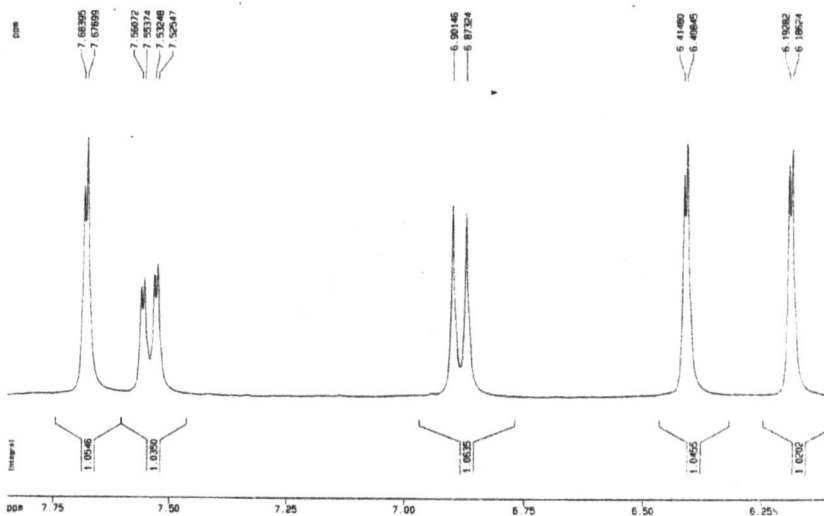

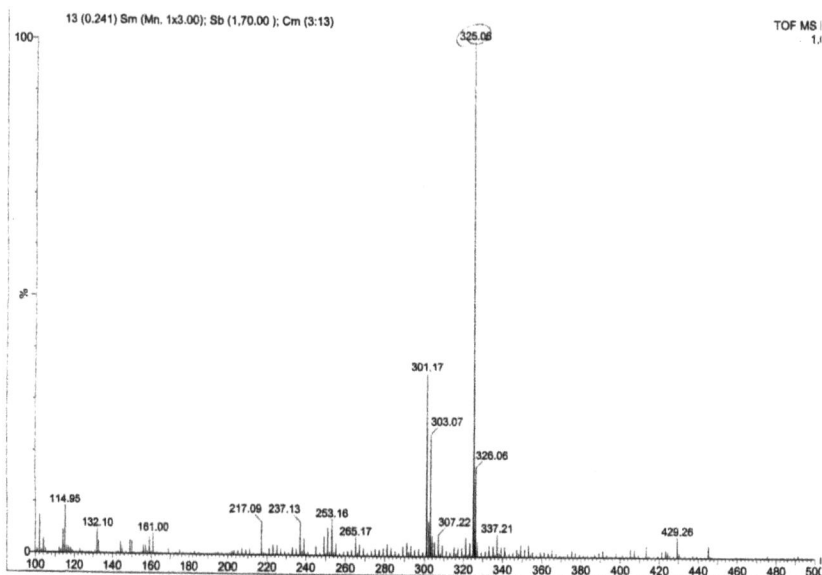

13 (0.241) Sm (Mn. 1x3.00); Sb (1,70.00); Cm (3:13)   TOF MS

Spectral Analysis:

IR (KBr, Cm^{-1}): 1666 (Carbonyl), 3409 (hydroxyl).

H^{1}NMR (DMSO-d6, 300 MHz): δ = 6.186-6.193 (d, 1H, J = 1.98 Hz), 6.408-6.414 (d, 1H, J = 1.92 Hz), 6.873-6.901 (d, 1H, J = 8.5 Hz), 7.525-7.560 (dd, 1H, J = 10.6 Hz), 7.676-7.683 (d, 1H, J = 2.0 Hz), 9.380 (s, 3H), 10.761-10.860 (t, 1H, J = 29.7 Hz), 12.498 (s, 1H).

ESIMS: m/z 302 [M$^+$], 325 [M + Na].

In structure 2, there should be 5 meta couplings. However spectra shows only 3 meta couplings. Further doublet of doublets at δ 7.525-7.560 ppm is possible only when there is ortho coupling along with meta coupling which is not present in the structure. Thus this structure was rejected.

In structure 3, it should give 2 singlets for the protons present on benzene ring of coumarin. But the spectrum is showing doublets. Based on this fact the structure can be rejected.

Structure 1 shows doublet of doublet and 3 meta couplings (shown by less J values ie < 2.0 Hz), which are present in the spectrum. Thus arrangements of hydroxyl groups seems to be more appropriate. Hence structure 1 is confirmed.

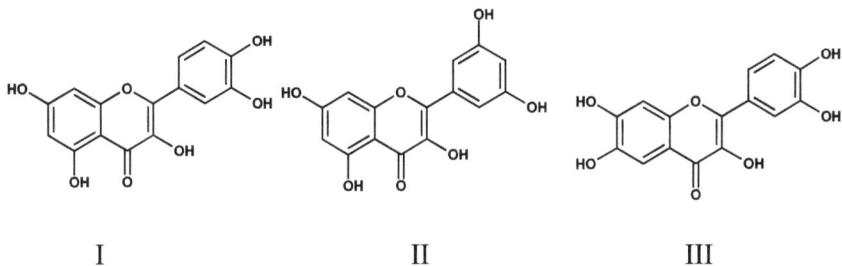

I   II   III

Part G

Essentials of Bioinformatics and Molecular Modeling

1

Rigid Docking Software: Hex

Introduction

Hex is an interactive protein docking and molecular superposition program, written by Dave Ritchie at the university of Alerdeen. Hex can run on most windows-XP, linux and Mac OS X Pcs. Hex does not work on intel mac pcs. Hex understands protein and DNA structures in protein data bank (PDB) format. Hex version 5-0 can also read small molecule SDF files. Hex is used for rigid docking between the receptor and the ligand for getting the best optimal fit. Hex provides high performance algorithms for structural systems biology.

URL- http:// www.loria.Fr/ □ritchied/hex

Docking Molecules

In order to run a docking process in Hex, there is need to load a receptor, ligand and a ligand pdb structure using the file pull- down menu.

Editing Origins

Molecular centroid is edited using the edit origins button (2 cross and a line) on the left hand border. It helps to activate the intramolecular axis, in order to view the new and old positions of the centroids. On reverting to the default pointer mode, the new origin is activated. The new origin only applies to closest residue atom of the desired pocket.

Editing Orientations

To control the molecular orientations, Hex uses the orientation control panel to select the ligand, and then to rotate the selected molecule using the Euler angle Alpha, Beta, and Gamma sliders. The intermolecular separation can be modified with the R slider at the bottom of the main window. Select '*commit*' after completing the new transformations into the molecule.

Docking and Molecular Modeling

1. Open the docking control panel and select '*activate*' to start docking operation. The steric scan (N=16) phase of the docking calculation will be performed at (1+40)/ 0.75 = 53 intermolecular separations, in +/- steps of 0.75 Å starting from the current distance posted in the R slider in the bottom border of the main window. The final search (N=25) phase will be applied to the highest scoring scan orientation in step of 0.75/2 Å.

2. In matching, Hex superposes or matches pairs of molecular structures using the same 3D density; representation as for docking. Superposition is very much like docking except now the search is for maximum similarity rather than maximum complementarity. Superposition numbers are calculated using the vander waals density of each molecule. However, this can be changed with the search mode selector to match shapes using both the exterior (vander waals) and interior (surface skin) density functions. Superposition calculations are controlled by the parameters in matching control panel. To calculate a superposition, use:

Controls ⟶ Matching ⟶ Activate

Saving Docking Results

'Compact output method' is to write a separate PDB file for each docking orientation. The current docking orientation can be written to a single PDB file by selecting:

File ⟶ Save ⟶ Both

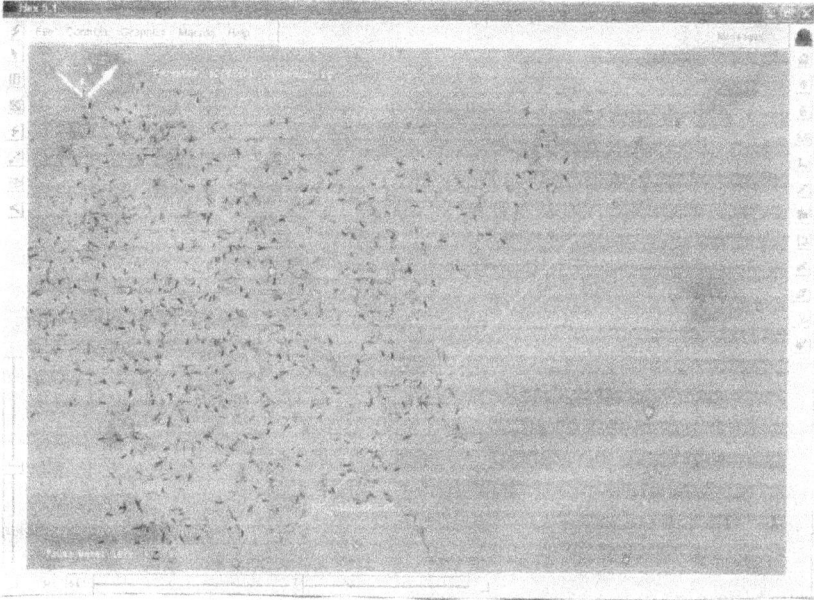

Fig. 1 Receptor with ligand in HEX.

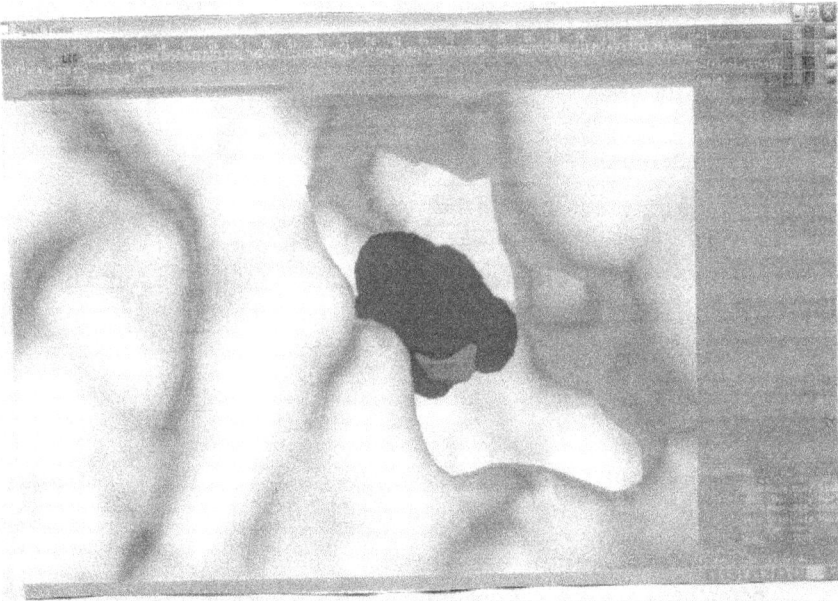

Fig. 2 HEX result in Pymol

2

To Run Ligand Building Software: Ligbuilder

Introduction

Ligbuilder is a general purposed program package. It was developed by **Dr. Renxiao Wang** at **Peking University of China**. Ligbuilder is written in ANSI C++ and has been tested on UNIX and LINUX platforms. It is written for **structure–based drug design** procedure which is based on the 3D structure of the target protein. It can automatically build ligand molecules within the binding pocket.

URL – http://www.chem.ac.vu/Chemistry/Soft/L/GBUILD.en.html

Features of Ligbuilder

1. The program analyzes the binding pockets of the target protein and derives the key interaction sites. A pharmacophore model is suggested & it could be applied to 3D database search for finding novel ligand molecules.

2. Molecules are constructed by using fragments as building blocks. Various kinds of structural manipulation are provided, such as growing, linking & mutation. Minimization of conformation is performed during the building up procedure while the target protein is kept rigid. Flexibility of the ligand molecules is considered.

3. User can choose either growing strategy or linking strategy to develop ligands.

4. Molecules are evolved by genetic algorithm. The fitness score of a molecule is evaluated by considering its chemical viability as well as binding affinity.

5. Chemical rules are adopted for evaluating "drug likeness" of the resultant molecules. Chemical stability, synthesis feasibility &

346

toxicity can also be taken into account by defining "forbidden structure" libraries.

6. All the input & output molecules are in popular format i.e protein in PDB format & ligands in Sybyl Mol.2 format.

Running Ligbuilder

Ligbuilder is used to build ligand molecule inside the binding pocket by growing the small ligand into a large molecule. Ligbuilder can follow 2 paths:

1. Pocket   Grow $\longrightarrow$ Process $\longrightarrow$
2. Pocket   Linking $\longrightarrow$ Process $\longrightarrow$

Input files for ligbuilder are: Hex result in pdb format and ligand in .mol2 format. Before giving Hex output file as input, 'CONNECT' information of ligand was added to the Hex file.

Ligbuilder is run in 3 steps:

1. Running Pocket

2. Running Grow

3. Running Process

Results

After completing its generation, it gives 10 derivatives of ligand in .mol2 format which are then converted into .pdb format and .mol format.

INPUT FILES

```
#
# input files
#
RECEPTOR_FILE              /root/Desktop/joita/pocket/jhe1.pdb
LIGAND_FILE                /root/Desktop/joita/pocket/lig.mol2
PARAMETER_DIRECTORY        ../parameter/
#
# output files
#
POCKET_ATOM_FILE           /root/Desktop/joita/pocket/joy_pocket.txt
POCKET_GRID_FILE           /root/Desktop/joita/pocket/joy_grid.txt
#
# key interaction sites and pharmacophore
#
KEY_SITE_FILE              /root/Desktop/joita/pocket/joy_key_site.pdb
PHARMACOPHORE_TXT_FILE     /root/Desktop/joita/pocket/joy_pharmacophore.txt
PHARMACOPHORE_PDB_FILE     /root/Desktop/joita/pocket/joy_pharmacophore.pdb
MINIMAL_FEATURE_DISTANCE   3.50
MAXIMAL_FEATURE_NUMBER     8
#
#
```

POCKET.INDEX FILE

```
#
# input files
#
SEED_LIGAND_FILE                    /root/Desktop/joita/grow/lig.mol2
POCKET_ATOM_FILE                    /root/Desktop/joita/grow/joy_pocket.txt
POCKET_GRID_FILE                    /root/Desktop/joita/grow/joy_grid.txt
#
# force field directory
#
PARAMETER_DIRECTORY                       ../parameter/
#
# fragment libraries
#
BUILDING_BLOCK_LIBRARY                 ../fragment.mdb/
FORBIDDEN_STRUCTURE_LIBRARY            ../forbidden.mdb/
TOXIC_STRUCTURE_LIBRARY               ../toxicity.mdb/
#
# structural construction parameters
#
GROWING_PROBABILITY                    1.00
LINKING_PROBABILITY                    0.50
MUTATION_PROBABILITY                   0.50
#
# chemical viability rules
#
APPLY_CHEMICAL_RULES                   YES
APPLY_FORBIDDEN_STRUCTURE_CHECK              YES
APPLY_TOXIC_STRUCTURE_CHECK            YES
MAXIMAL_MOLECULAR_WEIGHT               500
MINIMAL_MOLECULAR_WEIGHT              50
MAXIMAL_LOGP                          5.00
MINIMAL_LOGP                          -5.00
MAXIMAL_HB_DONOR_ATOM                 6
MINIMAL_HB_DONOR_ATOM                 2
MAXIMAL_HB_ACCEPTOR_ATOM              6
MINIMAL_HB_ACCEPTOR_ATOM              2
MAXIMAL_PKD                      10.00
MINIMAL_PKD                      5.00
#
# genetic algorithm parameters
#
NUMBER_OF_GENERATION                   20
NUMBER_OF_POPULATION                   3000
NUMBER_OF_PARENTS                300
ELITISM_RATIO                          0.10
SIMILARITY_CUTOFF                 0.90
#
# output files
#
POPULATION_RECORD_FILE               /root/Desktop/joita/grow/population.lig
LIGAND_COLLECTION_FILE               /root/Desktop/joita/grow/ligands.lig
#
```

GROW.INDEX FILE

```
#
# input files
#
LIGAND_COLLECTION_FILE               /root/Desktop/joita/process/ligands.lig
#
# chemical rules
#
MAXIMAL_MOLECULAR_WEIGHT              500
MINIMAL_MOLECULAR_WEIGHT             50
MAXIMAL_LOGP                         5.00
MINIMAL_LOGP                         -5.00
MAXIMAL_PKD                     10.00
MINIMAL_PKD                     5.00
#
# similarity threshhold
#
SIMILARITY_CUTOFF                0.90
#
# output files
#
NUMBER_OF_OUTPUT_MOLECULES           10
OUTPUT_DIRECTORY                /root/Desktop/joita/process/results.mdb/
#
```

PROCESS.INDEX FILE

3

To Run Flexible Docking Software: Autodock

Introduction

Autodock is a suite of automated docking tool. It is the product of **Scripps Research Institute**. It runs with the help of MGL tools (Python). It is designed to predict how small molecules, such as substrates or drug candidates, bind to a receptor of known 3D structure. It is also used in visualizing atomic affinity grids. They have also developed graphical user interface called **Autodock Tools** or **ADT** for short, which amongst other things helps to set up the bonds which will be treated as rotatable in the ligand and to analyze docking.

URL: http://autodock.scripps.edu/

Autodock actually consists of 2 main programs:

1. Autodock performs the docking of the ligand to a set of grids describing the target protein.
2. Auto-grid pre-calculates these grids.

Autodock calculations are performed in several steps:

1. Preparation of co-ordinate files using AUTODOCK TOOLS.
2. Precalculation of atomic affinities using AUTOGRID.
3. Docking of ligands using AUTODOCK.
4. Analysis of results using AUTODOCK TOOLS.

Molecular Docking using Autodock & Quantum

Autodock performs the docking of the ligand to a set of grids describing the target protein. Input files required are: receptor & ligand in .pdb format. Ligand is the output of ligbuilder. It is used to convert .pdb format into .pdbqt format.

Firstly receptor is converted from .pdb format to .pdbqt format. Secondly thin grid is set by building grid bon around the centre atom & then ligand is converted into .pdbqt format by making only 5 bonds rotatable.

After that a CONFIG file was generated which include all information about the receptor file, ligand file & grid box, it was then run in the command prompt.

Applications

1. X-ray crystallography
2. Structure based drug design.
3. Lead optimization
4. Virtual screening
5. Combinatorial library design
6. Protein-protein docking
7. Chemical mechanism studies.

Results

As a result, 10 output files were obtained. This can help for example, to guide organic synthetic chemist to design better binders.

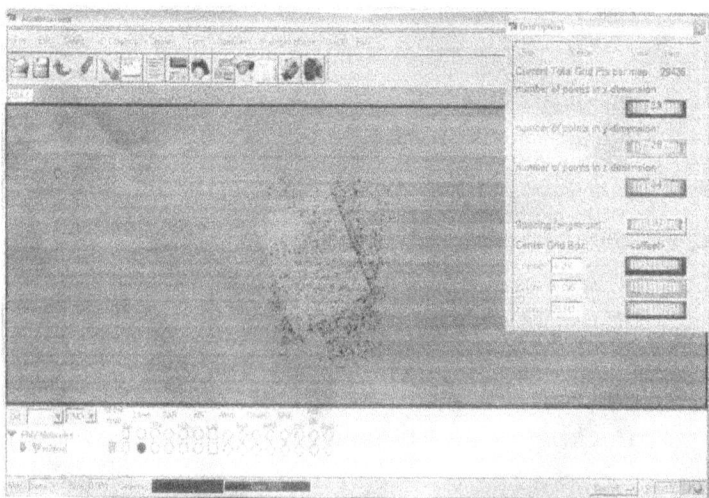

Fig. 3 Grid set in AUTODOCK

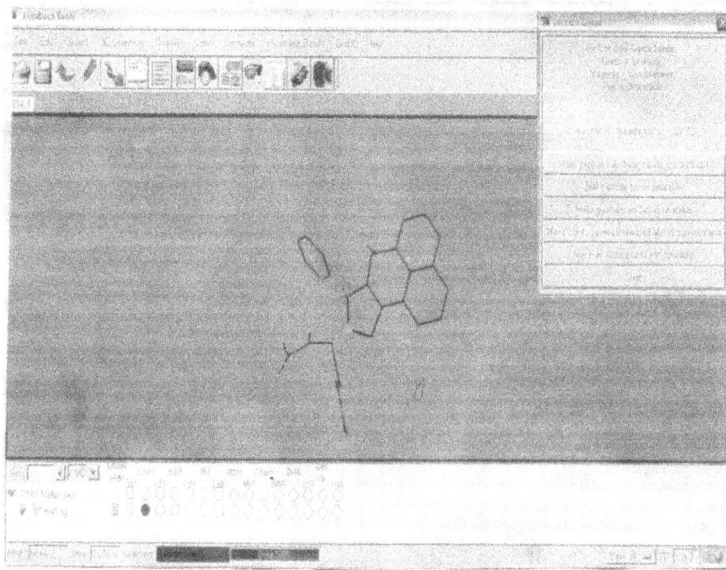

Fig. 4 Rotatable Bond

4

Protein Modelling Tool: Modeller

Introduction

Modeller is a computer program that models 3 dimensional structures of proteins & their assemblies by nullifying the spatial restraints. Modeller was originally written by **Andry Bali** & is currently maintained by **Ben Webb** at the **University of California, San Fransisco**. Modeller is most frequently used for homology or comparative protein structure modelling. Modeller is written in fortran- 90.

URL- http:// salilab.Org/ modeller/

Modeller is used in comparative modelling, in which the input files are Protein Data Bank (PDB) atom file of known protein structures & their alignment with the target sequence to be modelled & the output is a model of the target that includes all non-hydrogen atoms.

Preparing Input Files:

There are 3 kind of input files:

1. Atom coordinate file (.atm)
2. Alignment file (.ali)
3. Python file (.py)

Atom Files:

Each atom file is named code .atm where code is a short protein code, preferably the PDB code. We can also use file extensions .pdb & .ent instead of .atm. The code must be used as that protein's identifier throughout the modelling.

Alignment File:

One of the formats for the alignment file is related to the PIR database formats; this is the preferred format for comparative modelling. Influence of the alignment on the quality of the model cannot be overemphasized. At the aligned regions, modeller tries to derive a 3D model for the target sequence that is as close to one or the other of the template structures as possible & also satisfying stereochemical restraints (eg bond length, angles, nonbonded contacts), the inserted regions which do not have any equivalent segments in any of the templates, are modelled in the context of the whole molecule, using their sequence alone.

Script File:

Modeller is a command line tool and has no graphical user interface, instead it must be provided with a script file containing modeller commands. This is an ordinary python script.

INPUT FILES FOR MODELLER:

ATOM FILE (.atm)

ALIGNMENT FILE (.ali)

```
                       joy.py
1     from modeller.automodel import*
2     log.verbose()
      env=environ()
4     env.io.atom_files_directory='./:../joy.atm'
5     a=automodel(env,
6     alnfile='joy.ali',
7     knowns='joy',
8     sequence='joy1')
9     a.starting_model=1
10    a.ending_model=5
11    a.make()
```

PYTHON FILE (.py)

Running Modeller:

To run modeller with the script file:

1. Open command lines prompt on windows & click on the 'modeller' link on start menu & open windows command prompt. Set up to run MODELLER.

2. Change to the directory containing the atom, script & alignment files. Run MODELLER itself by typing the following at the command prompt- **Mod9v5 model-default.py** (Python file name)

3. Number of intermediate files are created as the program proceeds.

4. The final model is written to file 1fdx, b99990001.pdb.

5. Examine the model **default.log file** for information about the ru

5

To use Command Line Interface of Rasmol

Introduction

Rasmol is a molecular graphics program intended for the visualization of proteins, nucleic acids and small molecules. The original RASMOL manual was created by Roger Sayle. Rasmol runs on wide range of architecture and operating systems including Microsoft Windows, Apple Macintosh, UNIX & VMS systems. The program reads a molecule coordinate file & interactively displays the molecule on the screen in a variety of color schemes & molecule representations.

URL- http://www.openrasmol.org/doc/rasmol.html

Rasmol

Supported input file formats include PDB, Tripos Associate Alchemy & Sybyl Mol 2 formats, Molecular Design Limited (MDL) Mol File Format, Minnesota Supercomputer Centre (MSC) XYZ (XMOL) Format, CHARMm Format, CIF Format & mmCIF Format files. The loaded molecules can be shown as wireframe bonds, space filling spheres, ribbons, hydrogen bonding & dot surface representations. Rasmol displays 2 windows, the main graphics or canvas window with a black background & a command line or terminal window.

Command Line Interface:

Rasmol allows the execution of interactive commands typed at the "RASMOL>" prompt in terminal window, characters typed into either the terminal or display window are processed on command line. Keywords are case sensitive. All whitespace characters are ignored except to separate the keyword & the arguments of a command. Command lines are limited to a maximum of 256 characters. Placing a hash (#) character anywhere outside quotes terminates the line.

The commands/ keywords currently recognized by rasmol are:

1. Background – to set the color of "canvas" **Rasmol> Background <Color>**

2. Spacefill – to represent all of the currently selected atoms as solid sphere. **Rasmol> Spacefill**

   Spacefill off – turn off the representation of the selected atom as sphere. **Rasmol > Spacefill off**

3. Wireframe –

   Rasmol> Wireframe <Value>

   Represent each bond within the selected region of molecule as a cylinder a line or a depth cued vector. Display is turned on by the command "wireframe" or "wireframe on".

4. Select – **Rasmol > Select {<Expression>}**

   All subsequent commands for manipulation or modification of molecule only affect the currently selected region. The behaviour of the "select" command without any parameter is determined by the RASMOL 'hetero' & 'hydrogen parameters'.

5. Ribbons – **Rasmol > Ribbons**

   Displays a smooth solid "ribbon" surface passing along the backbone of the protein.

6. Color – **Rasmol > Color <Color>**

   Colors the selected region

7. CPK colors – **Rasmol > Color CPK**

   Developed by Corey, Pauling & later improved by Kultun. (carbon- light grey; oxygen- red, hydrogen- white, nitrogen-blue, sulphur- yellow.)

8. Reset – **Rasmol > Reset**

   Restores the original viewing transformation & centre of rotation.

9. Set picking – **Rasmol > Set Picking Centre.**

    Affects user interaction with a molecule displayed on the screen in rasmol.

10. Restrict – **Rasmol > Restrict {<expression>}**

    Define the currently selected region of the molecule & disables the most of those parts that are not selected.

11. SSBonds – **Rasmol >SSBonds**

    Represents the disulphide bridges of the protein molecule

12. H bonds – **Rasmol > H bonds**

    Represents the hydrogen bonding of the protein molecules backbone.

13. Label – **Rasmol > label**

14. Stereo – **Rasmol > Stereo on**

Provides side by side stereo display of images.

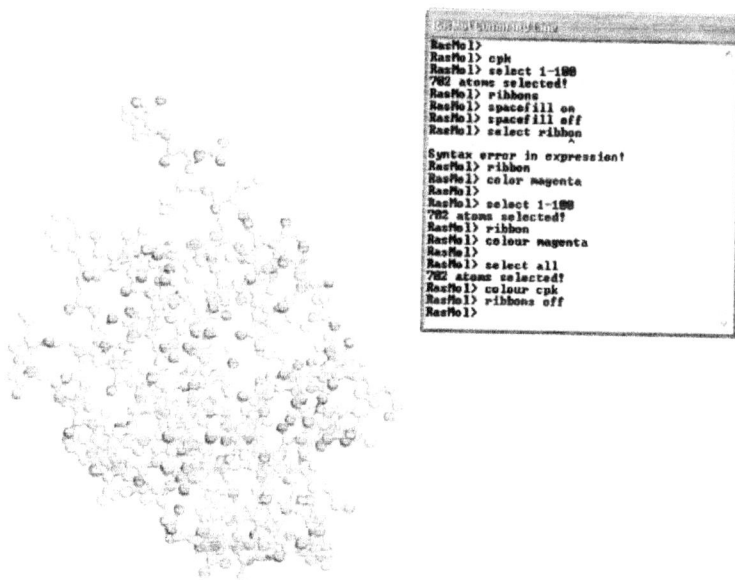

Fig. 5 Myoglobin with command line

```
RasMol Command Line
782 atoms selected?
RasMol> ribbons
RasMol> spacefill on
RasMol> spacefill off
RasMol> select ribbon
            ^
Syntax error in expression?
RasMol> ribbon
RasMol> color magenta
RasMol>
RasMol> select 1-100
782 atoms selected?
RasMol> ribbon
RasMol> colour magenta
RasMol>
RasMol> select all
782 atoms selected?
RasMol> colour cpk
RasMol> ribbons off
RasMol> reset
RasMol> restrict 15-75
415 atoms selected?
RasMol>
```

6

To Learn using
A Visualization Tool : Pymol

Introduction

Pymol is an open source, user sponsored, molecular visualization system created by Wairen Lyford Delano & commercialized by Delano Scientific LLC ,which is a private software company dedicated for creating useful tools that become universally accessible to scienfitic & educational communities. It is well suited for producing high quality 3D protein structures available in the scientific literature. PYMOL is one of few open source visualization tools available for use in structural biology. The **py** portion of the software's name refers to the fact that it extends & is extensible by the python programming language.

Procedure

The following steps were adopted to explore the visualization tool PYMOL & to manipulate the structure by selecting different commands provided on the PYMOL interface:

1. Firstly the structures of one of the molecule containing hetero group or porphyrin ring like MYOGLOBIN was downloaded from the PDB having ID 3HC9.

2. Myoglobin – It is a protein in humans which is encoded by the MB gene. Myoglobin is a single chain globular protein of 153 & 154 amino acids containing a heme prosthetic group in the centre around which the remaining apoprotein folds. It has 8 alpha helices & a hydrophobic core. It has a molecular weight of 16700 Daltons & is the primary oxygen carrying pigment of muscle tissues.

3. After that the structure was uploaded into the PYMOL through the PATH:

FILE > OPEN> PDB FILE > 3HC9

4. After opening the structure into PYMOL, the command prompt & the PYMOL structure visualization interface were opened side by side & the changes in the structures were viewed according to commands typed on the PYMOL command prompt or by selecting them directly on the PYMOL interface.

Results

The structure was viewed in different ways as described below:

PICK ATOM (PKAT) – To select particular atom on the structure which can be selected by the mouse shortcuts provided on the right corner at the bottom of the PYMOL interface as CTRL + MIDDLE BUTTON (MOUSE) & after that the desired atom can be selected by directly clicking on it in the structure viewed in PYMOL visualization interface.

- PICK TORSION (PTKB) – to select the torsion bond with 4 atoms selected through the mouse shortcuts Ctrl + Right Click (mouse)
- LABELS – label molecule according to chain, residue name, atom name.

L > LABEL > CHAINS

- COLORS – C > COLOR > REDS > REP
- CARTOON & STICK VIEW:

  PYMOL > show sticks

  PYMOL > show cartoon

  PYMOL > hide sticks / cartoon

- CLIP VIEW – The view provided the picture of inner atoms of the structure by rubbing the superficial atoms of the structure & was achieved by Ctrl + Shift +Right Click
- MEASUREMENTS – To measure distance between the 2 atoms by selecting them one by one
- BUILD FRAGMENT – To build different types of small fragments like bromine, iodine, acetylene, cyclobutane etc. In structure.

  BUILD > FRAGMENT > CYCLOBUTANE in the PYMOL prompt.

Molecule represented in ribbons with white background

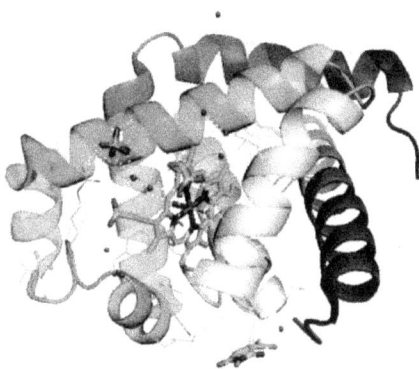

Molecule represented in cartoons with preset ligands

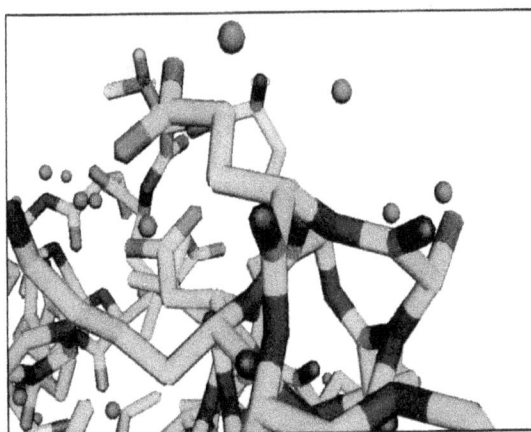

Molecule labeled with B-factor, VMD radius and occupacy

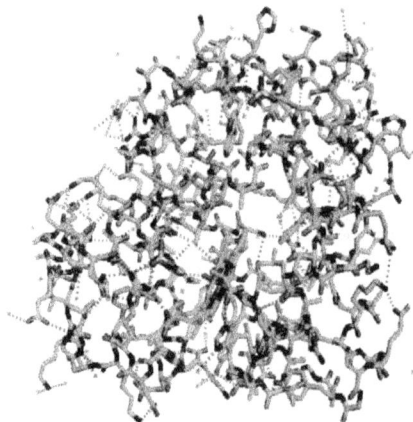

Molecule represented with polar contacts

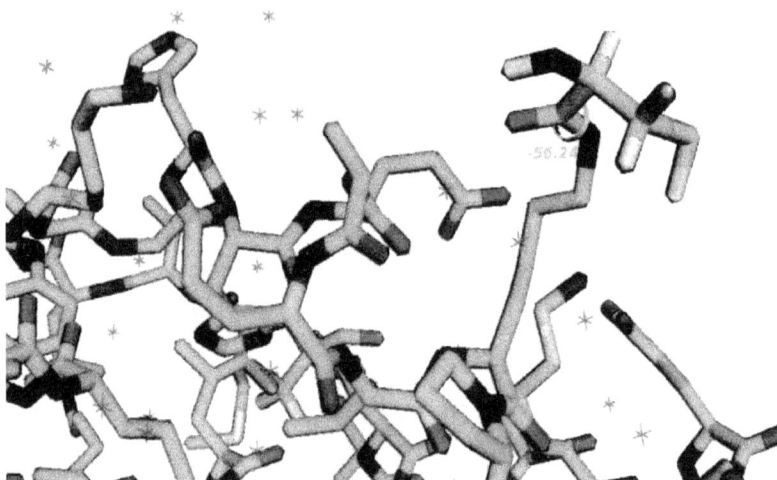

Fragment of residue added in the molecule

7

To Learn using A Visualization Tool : SPDBV

Introduction

Swiss PDB viewer has been developed since 1994 by Nicolas. SPDBV is tightly linked to SWISS MODEL, an automated homology modelling server developed within Swiss Institute of Bioinformatics (SIB) at the structural bioinformatics group at the Biozentrum Basel. Swiss PDB viewer is an application that provides a user friendly interface allowing analyzing several proteins at the same time. The proteins can be superimposed in order to deduce structural alignments & compare their active sites or any other relevant parts. Amino acid mutations, H-bonds, angles & distances between atoms are easy to obtain due to intuitive graphic & menu interface of SPDBV. Swiss PDB viewer can also read electron density maps & provides various tools to build into the density. In addition, various modelling tools are integrated and command files for popular energy minimization packages can be generated. The SPDBV also has an "import" utility that allows direct connection to the SPDBV server thereby providing direct access to database & powerful computing facilities.

Procedure

1. Structure of one molecule like **Myoglobin** was downloaded from the structural database PDB having PDB ID 3HC9
2. Upload structure to SPDBV through **PATH: SPDBV > FILE >OPEN > PDB FILE >3HC9**

Results

Structure was viewed in different ways as described below:

1. Labels: label its residue name, atom name,

   a.   **SPDBV >DISPLAY >LABEL> GROUPNAME**

2. Helical or Strand View: Select all residue from the 'control panel' option of the ' window' menu & then selecting 'helical' or 'strand' view from the option 'select omega, phi, psi' under the EDIT menu of SPDBV.

3. Computing Hydrogen Atoms: By selecting the option 'compute H-bonds 'under the display menu of SPDBV, the same tab also provided the option 'compute H-bonds distances' to find out the bond lenghs between hydrogen atoms.

4. Ramachandran Plot: Press Ctrl +R to view allowed & disallowed region.

5. Alignment & the force field energy for protein MYOGLOBIN was viewed by selecting the option "alignment" under WINDOW tab of SPDBV.

6. Electrostatic potential for MYOGLOBIN was viewed by selecting option "COMPUTE ELECTROSTATIC POTENTIAL" under the tool tab of SPDBV which showed the region above electrostatic potential in BLUE color & region below electrostatic potential in red color.

7. Molecular surface: By selecting option "COMPUTE MOLECULAR SURFACE" under TOOL tab of SPDBV.

8. All commands generated different types of manipulations in the structure of myoglobin which were observed by rotating, zooming & flipping the molecule in the visualization tool of SPDBV.

Molecule represented with bond length, bond angle, residue name and pair potential energy

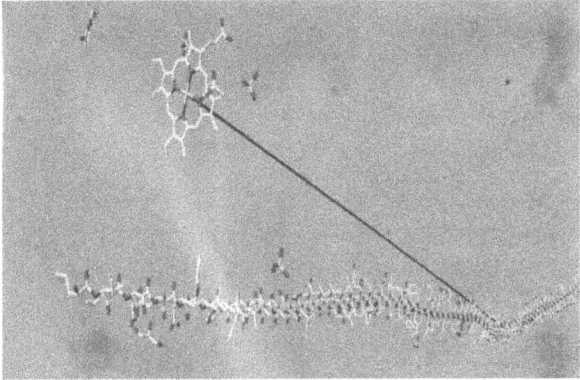

Molecule represented in beta strand structure

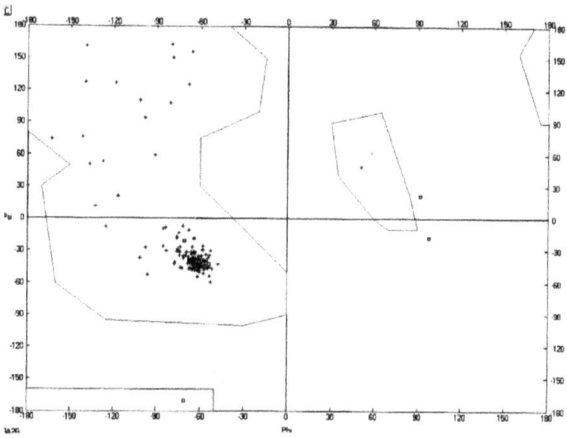

Ramachandran plot of the molecule

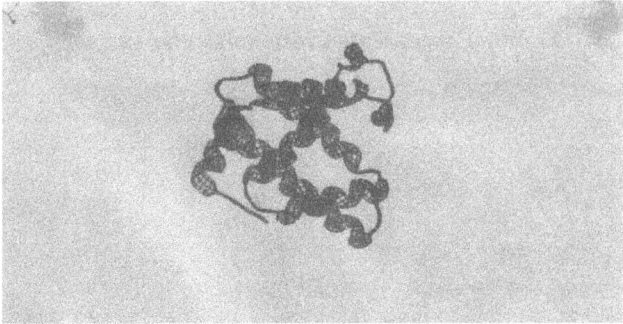

Molecule represented in ribbons

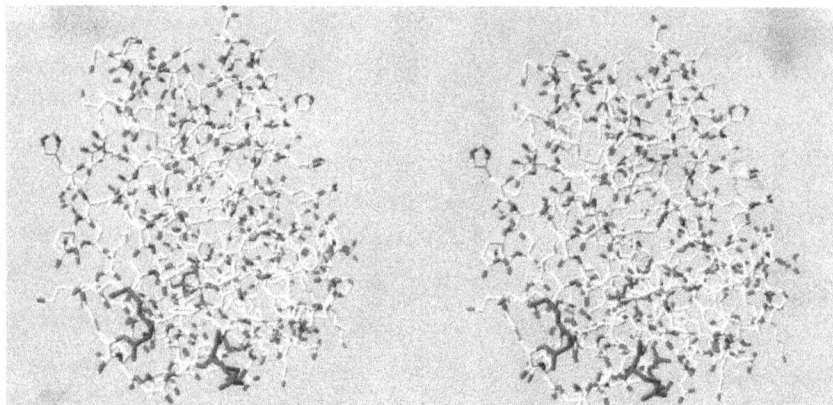

Stereo image the molecule

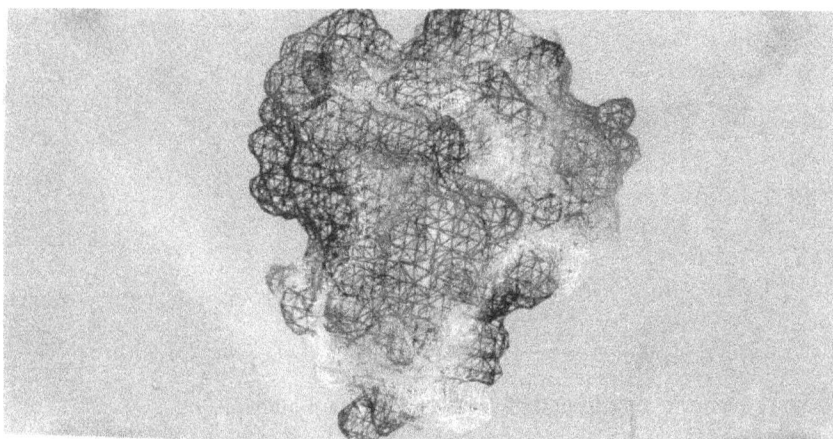

Molecule represented with molecular surface

Index

www.ingramcontent.com/pod-product-compliance
Lightning Source LLC
Chambersburg PA
CBHW050521190326
41458CB00005B/1618